Iron Phosphate Materials as Cathodes for Lithium Batteries

Pier Paolo Prosini

Iron Phosphate Materials as Cathodes for Lithium Batteries

The Use of Environmentally Friendly Iron in Lithium Batteries

Pier Paolo Prosini
Renewable Technical Unit, C.R. Casaccia
ENEA
via Anguillarese
00123 Rome
Italy
e-mail: pierpaolo.prosini@enea.it

ISBN 978-1-4471-6075-5 ISBN 978-0-85729-745-7 (eBook)
DOI 10.1007/978-0-85729-745-7
Springer London Dordrecht Heidelberg New York

British Library Cataloguing in Publication Data
A catalogue record for this book is available from the British Library

Cover design: eStudio Calamar S.L.

Printed on acid-free paper

Springer is part of Springer Science+Business Media (www.springer.com)

Preface

In the book, our efforts to make lithium iron phosphate ($LiFePO_4$) a suitable material for lithium-ion batteries are presented. It was found that carbon, added before the formation of the crystalline phase, was effective on improving the electrochemical properties of the material in terms of practical capacity and charge/discharge rate. The full capacity (170 Ah kg^{-1}) was attained when discharging the cell at 80°C and C/10 rate. To evaluate the lithium chemical diffusion the lithium insertion in $LiFePO_4$ was treated with a Frumkin-type sorption isotherm. The diffusion coefficient was found lower than the theoretical value of seven orders of magnitude. The poor electrochemical performance exhibited from the material was related to the relatively low value of the calculated diffusion coefficient. The reduction of the grain size was supposed to be one of the possible routes to enhance the performance of $LiFePO_4$. Solution-based, low-temperature approaches can access metastable phases and unusual valence states that are otherwise inaccessible by conventional solid-state reactions. Amorphous $FePO_4$ was prepared by sol-gel precipitation followed by air oxidation. Amorphous $FePO_4$ was also prepared by spontaneous precipitation from equimolar aqueous solutions of iron and phosphate ions using hydrogen peroxide as an oxidizing agent. The material was able to reversibly intercalate lithium. Amorphous $FePO_4$ was lithiated to obtain amorphous $LiFePO_4$. Nanocrystalline $LiFePO_4$ was prepared by thermal treatment of the amorphous material. Electrochemical tests have been done to evaluate factors affecting rate performance and long-term cyclability of the material. It was shown that nanocrystalline $LiFePO_4$ showed good electrochemical performance both at low- and high-discharge rate. At C/10 discharge rate the material delivered a specific energy close to the theoretical one. To test both the effects of carbon coating and grain size reduction on the same sample, carbon covered nanocrystalline $LiFePO_4$ was prepared. The so-obtained material showed the best electrochemical performance in terms of specific capacity, energy density, power density, and cyclability. A model to explain lithium insertion/extraction and predict the discharge curves at various rates was also illustrated.

October 2011 Pier Paolo Prosini

Acknowledgements

Part of the work was made in collaboration with Francesco Cardellini, Maria Carewska, Marida Lisi, Carla Minarini, Stefano Passerini, and Silvera Scaccia (ENEA) Elvira M. Bauer, Carlo Bellitto, Luciano Cianchi, Guido Righini, Gabriele Spina, Daniela Zane (CNR), Alessandro dell'Era, Mauro Pasquali (Un. of Roma "La Sapienza"), and Pawel Wisniewski (Un. of Warsaw) and I would like to thank the co-authors for their collaboration.

Contents

Chapter 1
Electrode Materials for Lithium-ion Batteries

1.1 Cathode Materials for Lithium-ion Batteries

Lithium-ion batteries represent the top of technology in electrical storage devices. Lithium-ion batteries with $LiCoO_2$ cathode and carbon anode were introduced by SONY in early 1990s [1]. High-energy density, high power, and long service life make lithium–ion batteries suitable for several applications from mobile phones to laptops and power tools. Energy densities of about 140–150 Wh kg^{-1} are now available in cells using a metal can, with higher values of 160–170 Wh kg^{-1} in cells with light-weight packaging [2]. The material that is currently used as a cathode in lithium–ion batteries is lithium cobaltite ($LiCoO_2$) which is a member of the $LiMO_2$ series (where M = V, Cr, Co and Ni). These compounds have a lamellar-type rock-salt structure, based on a compact network of oxygen atoms in which the lithium atoms and those of the transition metal (M) occupy ordered and alternating layers between the planes (Fig. 1.1). Lithium can be reversibly inserted and extracted at a fairly constant potential value of around 4.0 V versus Li. Even lithiated manganese oxide ($LiMn_2O_4$) with a spinel structure, in which the lithium occupies the tetrahedral sites and manganese the octahedral sites (Fig. 1.2), was proved able to reversibly intercalate lithium ions at a potential of about 4.1 V versus Li.

The theoretical energy density of $LiCoO_2$ and $LiNiO_2$ is about twice that of $LiMn_2O_4$, but in practice only half of the lithium content can be removed from the first two compounds without compromising their structural stability. As a consequence, the energy density for these three materials is of the same order of magnitude. Further significant increases in energy density are possible by the use of $LiCo_{1-x}Ni_xO_2$ [3] or $LiNi_{1/3}Co_{1/3}Mn_{1/3}O_2$ [4] instead of $LiCoO_2$. This raises the capacity to around 180 Ah kg^{-1}.

The choice of the cathode active material must be done not only in relation to battery performance, but also in relation to cost and safety. Due to high cost and material shortage large-capacity batteries, such as batteries for automotive applications, based on cobalt cathode are not realistic. Identified world cobalt resources are about 15 million tons. The Cobalt Development Institute estimated for 2010 a

P. P. Prosini, *Iron Phosphate Materials as Cathodes for Lithium Batteries*,
DOI: 10.1007/978-0-85729-745-7_1, © Springer-Verlag London Limited 2011

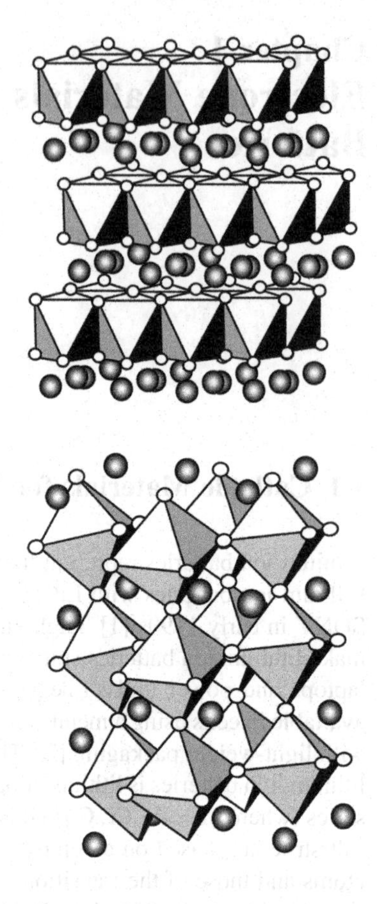

Fig. 1.1 Lamellar-type rock-salt structure of $LiMO_2$: ○ oxygen, ● lithium, [polyhedron] vanadium, chromium, cobalt, nickel

Fig. 1.2 Spinel structure of $LiMn_2O_4$: ○ oxygen, ● lithium, [polyhedron] manganese

world consumption of 57,000 tons of cobalt. The large cobalt consumption will make cobalt resources completely exhausted in less than 300 years. To store 40 kWh of energy, necessary to power a light electric vehicle, 250 kg of lithium–ion batteries are needed, of which about 80–100 kg is represented by the weight of the cathode. The cobalt weight in the cathode can vary from 15 kg (for the ternary $LiNi_{1/3}Co_{1/3}Mn_{1/3}O_2$) up to 50 kg (for pure $LiCoO_2$) per vehicle. By considering the use of 10% of all the cobalt extracted in 1 year to build batteries for electric vehicles and an average of 30 kg of cobalt per electric vehicle the number of vehicles that is possible to build is limited to 190,000 per year. Additionally, the price of $ 60.00 per kg appears to be too high for use in automobiles.

1.2 Iron-Based Materials for Lithium-ion Batteries

From many different aspects iron is a very attractive metal to build large-size batteries for powering electric vehicles or for realizing dispersed electrical power sources. Iron is very plentiful and environmentally benign; it is the fourth most abundant element in the earth's crust, being outranked only by aluminium, silicon, and oxygen. It has been the most important metal in the development of civilization to the present time. The earliest record of iron used by man dates back to at least 2000 BC [5]. Practically everything that surrounds us in our lives contains iron, or iron has been used in its manufacture.

However, for a series of reasons, iron and its derivatives have not met with success as electrode material for lithium–ion batteries. In fact, in the iron-based oxides containing O^{2-} as anion, the Fe^{4+}/Fe^{3+} redox energy tends to lie too far below the Fermi energy with respect to a lithium anode, while the Fe^{3+}/Fe^{2+} couple is too close to it.

Iron oxide (Fe_2O_3) was initially tested as a anode for lithium cells. Ohzuku et al. [6] tried to electrochemically insert lithium into Fe_2O_3. They found that two lithium equivalents can be introduced into the iron oxide and postulated the formation of FeO and Li_2O. In a following work, Di Pietro et al. [7] prepared Li_6FeO_3 by Fe_2O_3 electroreduction. Abraham et al. [8] prepared Li_xFeO_3 by lithiation of iron oxide with Li-naftalide in THF and tested the so obtained material as anode in lithium cells.

On the other hand, lithiated iron oxides ($LiFeO_2$) were tested as a cathode. $LiFeO_2$ synthesized by high temperature solid-state reaction, can crystallize in three polymorphic modifications: a disordered rock-salt structure, a tetragonal structure and an intermediate combination of the two structures [9]. $LiFeO_2$ with a rock-salt structure synthesized by H^+/Li^+ ionic-exchange was tested as a cathode for lithium–ion batteries and it was shown that lithium can be electrochemically removed from the structure [10]. $LiFeO_2$ iso-structural with α-$NaFeO_2$ was prepared by a Na^+/Li^+ ionic-exchange [11–13] and from a hydrothermal reaction between $FeCl_3$ and LiOH [14]. The electrochemical performance of the so obtained materials was not impressive. Corrugated $LiFeO_2$, iso-structural with $LiMnO_2$, was prepared by an ionic-exchange between γ-FeOOH and LiOH [15]. The material showed a specific capacity of about 100 Ah kg^{-1} and a capacity fade of 1.4% per cycle. The average discharge voltage was about 2.0 V versus Li. Kim and Manthiram prepared nanocrystalline lithium iron oxide and studied the lithium intercalation properties of the material. When synthesized by an oxidation reaction in solution followed by firing the precursors at 200°C, nanocrystalline iron oxide exhibits capacities as high as 140 Ah kg^{-1} with excellent cyclability over a wide voltage range [16].

Other compounds based on iron have been proposed as a cathode for lithium–ion battery. Among them Li_2FeS_2 [17], FeS_2 [18, 19], FeOCl [20], and $FeFl_3$ [21]. Almost all these compounds showed poor cyclability, high irreversibility, and low discharge voltage.

1.3 Iron Phosphates as Cathode Materials for Lithium-ion Batteries

The use of polyanions such as XO_n^{m-} (X = Mo, W, S, P, As) has been shown to lower the Fe^{3+}/Fe^{2+} redox energy to useful levels [22, 23]. Monoclinic and hexagonal iron sulphate ($Fe_2(SO_4)_3$) with a NASICON-related framework were

Table 1.1 Theoretical gravimetric specific energy (in Wh kg^{-1}) and volumetric energy density (in Wh l^{-1}) for various materials used as a cathode in lithium-ion batteries compared to $LiFePO_4$

Cathode material	$LiCoO_2$	$LiNiO_2$	$LiMn_2O_4$	$LiFePO_4$
Wh kg^{-1}	510	640	420	578
Wh l^{-1}	2600	3000	1700	2000

synthesized and tested as lithium intercalation hosts [23]. Both forms have been shown to give a flat discharge voltage of 3.6 V versus Li and a specific capacity of about 100 Ah kg^{-1}. The effect of the structure on the Fe^{3+}/Fe^{2+} redox energy of several iron phosphates was studied. $Li_3Fe_2(PO_4)_3$, $LiFeP_2O_7$, $Fe_4(P_2O_7)_3$ and $LiFePO_4$ were investigated by Padhi et al. [24]. Phosphates also include materials with high oxidation–reduction potential, such as $LiMnPO_4$ [25], $LiVPO_4F$ [26], $LiCoPO_4$ [27], and $LiNiPO_4$ [28] and multi-electron redox intercalation compounds, such as $Li_2NaV_2(PO_4)_3$ [29] and $Li_3V_2(PO_4)_3$ [30]. Among these, lithium iron phosphate ($LiFePO_4$) with an olivine structure was investigated as suitable positive-electrode material for rechargeable lithium–ion batteries [31].

Table 1.1 shows the theoretical gravimetric specific energy and volumetric energy density for various materials used as a cathode in lithium–ion batteries compared to $LiFePO_4$. $LiFePO_4$ has a practical specific energy as high as 578 Wh kg^{-1}. The specific energy observed for $LiFePO_4$ is very close to the theoretical specific energy predicted for $LiNiO_2$ and larger than the specific energy for $LiCoO_2$ and $LiMn_2O_4$. The low gravimetric density of the material (3.577 g cm^{-3}), however, is detrimental for volumetric energy density. The calculated volumetric energy density exceeds 2000 Wh l^{-1} which is lower than the energy density for $LiCoO_2$ and $LiNiO_2$, but higher than the energy density for $LiMn_2O_4$.

$LiFePO_4$ is a natural product known by the name of Triphylite. The first crystallographic characterization was made by Yakubovich [32] on a sample coming from Palermo Mine, New Hampshire, USA. The compound was correlated to the olivine group. The olivine can be considered the hexagonal structural analogue of spinel.

In the olivine structure it is possible to observe two octahedral sites (Fig. 1.3). Iron is located on octahedral sites and is separated by PO_4 bridges. Lithium ions occupy adjacent octahedral sites along the c-axes of the a-c planes. Lithium can be chemically extracted from $LiFePO_4$ thus leaving a new phase, iron phosphate ($FePO_4$), isostructural with heterosite ($Fe_{0.65}Mn_{0.35}PO_4$), with the same spatial group of $LiFePO_4$. During lithium extraction, the framework of the ordered olivine is retained with minor displacements [31]. Crystallographic parameters for $LiFePO_4$ and $FePO_4$ are reported in Table 1.2. The lithium extraction leads to a contraction of the *a*- and *b*-parameter and a small increase of the *c*-parameter. The cell volume decreases by about 6.8% and the density increases by about 2.59%.

Padhi et al. [31] showed that lithium can be electrochemically extracted from $LiFePO_4$ and inserted into $FePO_4$ along a flat potential at 3.5 V versus Li. Also the electrochemical lithium extraction proceeds via a two-phase process and the

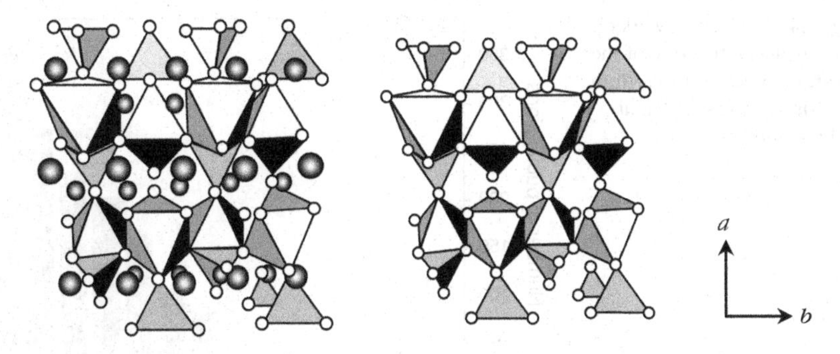

Fig. 1.3 Olivine structure of $LiFePO_4$ and $FePO_4$: ○ oxygen, ● lithium, △ phosphorus, ▲ iron

Table 1.2 The table reports the spatial group, the cell parameters, and the cell volume for $LiFePO_4$ and $FePO_4$

Material	$LiFePO_4$	$FePO_4$
Spatial group	Pb nm	Pb nm
a (Å)	6.008	5.792
b (Å)	10.334	9.821
c (Å)	4.693	4.788
Volume (Å^3)	291.392	273.357

$FePO_4$ framework of the ordered olivine $LiFePO_4$ is retained with minor adjustments. The $LiFePO_4$ charge and discharge process can be broadly described as:

$$LiFePO_4 \underset{\text{discharge}}{\overset{\text{charge}}{\longleftrightarrow}} Li_{(1-x)}FePO_4 + xLi^+ + xe^- \quad (1.1)$$

$LiFePO_4$ has a theoretical specific capacity of 170 Ah kg^{-1}. Nevertheless, the electrochemical insertion/extraction of lithium conducted at a specific current as low as 2.1 A kg^{-1} (C/81 rate), was limited to about 0.6 Li per formula unit. A "radial model" for the lithium motion was proposed to explain the poor electrochemical performance of the material [31].

The capacity exhibited from the material was strictly related to the current density used. The observation that the capacity was restored when reducing the discharge current indicates that the loss in capacity was a diffusion-limited phenomenon within a single grain. For this reason the poor electrochemical material utilization was associated with a diffusion-limited transfer of lithium ions across the two-phase interface. Galvanostatic intermittent titration technique (GITT) and impedance spectroscopy (IS) were used to calculate the diffusion coefficient of lithium (D_{Li}) in $Li_{(1-x)}FePO_4$ as a function of the lithium content. Although the theory of GITT and IS was shown to be strictly valid for solid solution reactions, reasonable values for D_{Li} can also be obtained for the case of two-phase reactions if the interaction between the intercalation sites are moderate. The calculated lithium–ion diffusion coefficient in $Li_{(1-x)}FePO_4$ is around 10^{-7}–10^{-8} cm^2 s^{-1} [33], while the actual measure indicates that the diffusion coefficient may be lower

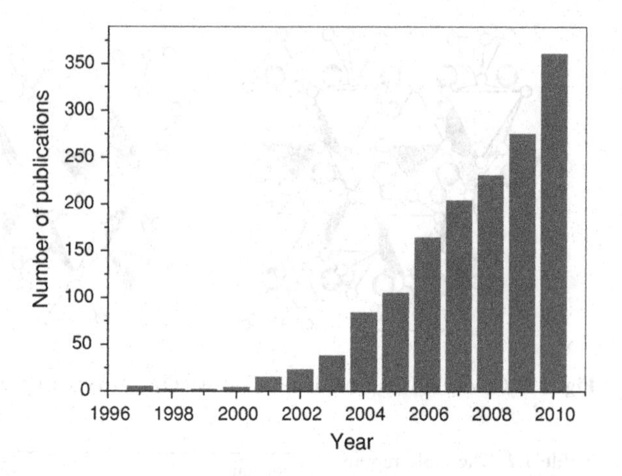

Fig. 1.4 Number of works published as a function of the publication year starting from the original work of Padhi and co-workers

than the theoretical value of 7 orders of magnitude (1.8×10^{-14} cm^2 s^{-1} [34]). The calculated D_{Li} as a function of x in carbon added $Li_{1-x}FePO_4$ was found to range from 1.8×10^{-14} to 2.2×10^{-16} cm^2 s^{-1} for $LiFePO_4$ and $FePO_4$, respectively, with a minimum corresponding to the peak in the differential capacity [35]. $LiFePO_4$ can be considered a mixed ionic/electronic conductor with an electronic band-gap of 0.3 eV. The electronic and ionic conductivities at room temperature are very low, about 10^{-9} S cm^{-1} and 10^{-5} S cm^{-1}, respectively. So, to make $LiFePO_4$ a suitable cathode material for lithium–ion batteries its electronic and ionic conductivity must be increased.

1.4 Methods to Improve the Electrochemical Performance of $LiFePO_4$

Since the first report in 1997 on the electrochemical properties of $LiFePO_4$ [31] the interest for the material progressively increased. Figure 1.4 shows the publication number as a function of the publication year starting from the work of Padhi and co-workers.

Starting from 1997 the number of publications continuously increased and more than 360 works on this topic have been published in 2010. In the literature it is possible to find several methods to improve the $LiFePO_4$ electrochemical properties. The main methods include: (i) ionic substitution, (ii) synthesis of particles with well-defined morphology, and (iii) synthesis of carbon covered material. All three of these methods have been explored and the results have been found very promising. The doping with metal ions was found effective to improve the electrochemical performance of the material. The result was related to the increase of the electric conductivity of the doped material. On the other hand, the enhanced electrochemical properties of $LiFePO_4$ nano particles were related to the grain size reduction and consequently to the decrease of the length of lithium ions channels inside the material. The dispersion of carbon particles in $LiFePO_4$ was also found effective to increase the electric contact between the $LiFePO_4$ grains enhancing the electron transfer and the electric conductivity of the composite electrode. The co-existence of several mechanisms of action makes the use of the different methods at the same time possible.

Metal doping was widely used to improve the electrical conductivity of $LiFePO_4$. Several metal ions were used to dope $LiFePO_4$, including Mg^{2+}, Ni^{2+}, Co^{2+}, Al^{3+}, Ti^{4+}, Zr^{4+}, Nb^{5+}, W^{6+}, etc. [36–39]. S-Y. Chung [36] found that Nb^{5+} and other metal ions in the 4α position increased the $LiFePO_4$ conductivity up to $3 \times 10^{-3} \sim ^{4}$ S cm^{-1}, even more than the $LiCoO_2$ ($\sim 10^{-3}$ S cm^{-1}) and $LiMn_2O_4$ (10^{-5} S cm^{-1}) conductivity.

However, there is considerable controversy about the effect of niobium doping. In fact while doping can significantly increase the electric conductivity of the material, the improvement in the electrochemical performance could be related to the formation of another conductive material inside the $Li_{(1-x)}Nb_xFePO_4$ phase. Nazar et al. [40] showed niobium does not significantly improve the conductivity of lithium iron phosphate by doping, but the synthesis process, particularly in high temperatures, brings to the formation of conductive iron phosphides (such as Fe_2P) formed on the surface of the material. The presence of a conductive metal phosphide nano-network can increase the electrochemical performance of the material by improving the conductivity of grain boundaries.

$LiFePO_4$ is a mixed ionic-electronic conductor in which both lithium ions and electrons could dominate the transport phenomena. The reduction of the grain size, and consequently the reduction of the diffusion length both for electrons or ions, is one of the possible routes to enhance the performance of the material. Several methods such as solid-state, sol–gel, hydrothermal, microwave, spray pyrolysis, precipitation, and emulsion drying have been used to synthesize $LiFePO_4$ [41–49]. Depending on the synthesis used for the preparation of the material, $LiFePO_4$ samples with different morphology and grain size have been obtained. The particle size is strictly related to the electrochemical performance. In 2001, Yamada et al. [50] proposed to reduce the particle size to overcome the low utilization of $LiFePO_4$. Andersson et al. [51] followed the electrochemical delithiation and subsequent lithiation of $FePO_4$ by in-situ X-ray diffraction and Mössbauer spectroscopy. They found that about 20–25% of $LiFePO_4$ remained unconverted and that this figure could be reduced by appropriate manipulation of particle-size and particle–surface morphology. In a further work [52], Andersson showed that the capacity during the first lithium extraction was higher than the capacity recovered during the following discharge cycles. Furthermore, the capacity increased with temperature, supporting the notion that the diffusion of lithium within each particle was the limiting step. In addition to the "radial model", a "mosaic model" was proposed to explain both the source of the first-cycle capacity loss as well as the poor electrochemical performance of the material [52]. Prosini et al. [53] prepared undoped nano-or sub-micron $LiFePO_4$ with a particle size of 100–150 nm. The material was discharged at 510 A kg^{-1} with a specific capacity of 140 Ah kg^{-1}. Masquelier et al. [54] found that the specific capacity at 850 A kg^{-1} discharge rate of undoped, carbon uncoated lithium iron phosphate with particle size of about 140 nm is 147 Ah kg^{-1}. Recently, Y.M. Chiang that initially proposed the niobium doping has begun to focus on nano-particles of lithium iron phosphate [55].

Carbon-coated lithium iron phosphate particles were also prepared to improve the electrochemical performance of the material. The carbon coating of lithium iron phosphate can not only improve the material electric conductivity, but can also effectively control the particle size. In 1999, Ravet et al. [56] proposed the use of an organic compound (sucrose) as a carbon source to prepare in situ modified carbon-coated material. It was found that the discharge capacity of 1% carbon-coated $LiFePO_4$ discharged at 170 A kg^{-1} was about 160 Ah kg^{-1}, very close to the theoretical value. Carbon-coated $LiFePO_4$ was also synthesized by heating the precursors in presence of high-surface area carbon-black [57]. SEM micrographs confirmed that the addition of fine carbon powder before the formation of the crystalline phase reduced the $LiFePO_4$ grain size (the average grain size was less than 10 μm). Electrochemical tests showed that the presence of carbon enhanced the electrochemical performance in terms of practical capacity and charge/discharge rate. Electrochemical tests were conducted at various temperatures. The specific capacity was seen to increase upon rising the cell temperature. The full theoretical capacity was recorded for a cell discharged at 80°C and 17 A kg^{-1} rate. In 2001 Nazar et al. [58] showed that nano-particles of carbon-coated $LiFePO_4$ can be discharged at 850 A kg^{-1} with a capacity of 120 Ah kg^{-1}. They claimed that carbon can also improve the electric and ionic conductivity of $LiFePO_4$. Chen and Dahn [59], tried a variety of ways to obtain carbon-coated materials with enhanced energy tap density. They stressed that the right way to obtain high-energy density carbon-coated material must take in consideration the material capacity expansion and the tap density effects. In 2003, Barker et al. [43] reported the use of "carbothermal reduction" (CTR) for the preparation of carbon coated lithium iron phosphate using carbon both as reducing agent and carbon source. Lithium iron phosphate produced by the CTR method showed discharge capacity up to 156 Ah kg^{-1}. The CTR method was claimed to be an effective method for large-scale synthesis. At the same time the carbon could be dispersed in tiny particles inside the $LiFePO_4$, increasing the electric conductivity and enhancing the lithium diffusion in the iron phosphate particles. CTR was considered not only the best way to achieve industrialization but also represented a way to optimize the material in terms of specific capacity and tap density. Zhu et al. [60] used polypropylene as an alternative source of inorganic carbon. The results of research show that carbon produced by polypropylene decomposition at high temperature is effective to prevent grain aggregation. The advantages of this approach are the following: the polymer decomposition can reduce the synthesis temperature and shorten the reaction time; the polymerization decomposition leads to the decentralization of carbon atoms in the reaction system; the formation of the carbon film reduces the particles size of lithium iron phosphate. Doeff et al. [61] studied the effect of different sources of carbon and the structure of the carbon surface layer on the electrochemical performance of $LiFePO_4$. It was found that carbon fibres and carbon nanotubes can significantly improve the performance of $LiFePO_4$.

They also found that the electrochemical performance of the composite material depends on the carbon structure rather than the carbon content. As the conductivity of sp^2 hybridized carbon graphite is higher than the conductivity of disordered sp^3 hybridized carbon, the electrochemical performance of carbon-covered $LiFePO_4$ improved with increasing sp^2/sp^3 ratio. Dominkó et al. [62], using a sol–gel technique prepared porous carbon-coated material with different carbon layer thicknesses. They evaluated that the increasing of the thickness of the carbon layer from 1.0 nm to 10.0 nm leads to an increase of carbon content of 3.2%. The specific discharge capacity recorded at 170 A kg^{-1} for 10 nm thick carbon coated $LiFePO_4$ was approximately 140 Ah kg^{-1}. It was claimed that both particle-size minimization and intimate carbon contact are necessary to optimize the electrochemical performance. A similar concept was proposed by Croce et al. [63] that dispersed fine copper or silver powder (0.1 μm average particle size) in the $LiFePO_4$ precursors before the formation of the crystalline phase. The metal particles were used as nucleation sites for the growth of $LiFePO_4$ particles as well as to enhance the overall electric conductivity. As a result, micrometric $LiFePO_4$ "metal added" powder was obtained.

NTT and SONY groups have investigated the effect of the synthesis temperatures on the electrochemical performance of $LiFePO_4$ [64, 65]. Both found that the reduction of the synthesis temperature decreased the particle size, increasing the specific surface area and enhancing the electrochemical performance. Yamada et al. [64] found a maximum in the specific capacity (160 Ah kg^{-1}) for the sample prepared at 550°C. The specific surface area was about 10 m^2 g^{-1} with a grain size ranging from 0.2–30 μm.

To summarize, the literature results clearly show that both particle-size reduction and intimate contact with the conductive binder are necessary to optimize the electrochemical performance of $LiFePO_4$-based electrodes. In the book our efforts to make $LiFePO_4$ a suitable cathode material for lithium–ion batteries are presented.

References

1. K. Sekai, H. Azuma, A. Omaru et al., Lithium-ion rechargeable cells with $LiCoO_2$ and carbon electrodes. J. Power Sour. **43**, 241–244 (1993)
2. B.A. Johnson, R.W. White, Characterization of commercially available lithium-ion batteries. J. Power Sour. **70**, 48–54 (1998)
3. R.J. Gummow, M.M. Thackeray, Lithium-cobalt-nickel-oxide cathode materials prepared at 400°C for rechargeable lithium batteries. Solid State lonics **53**, 681–687 (1992)
4. N. Yabuuchi, T. Ohzuku, Novel lithium insertion material of $LiCo_{1/3}Ni_{1/3}Mn_{1/3}O_2$ for advanced lithium-ion batteries. J. Power Sour. **119**, 171–174 (2003)
5. S.L. Goodale, *Chronology of Iron and Stell* (Penton Publishing Co, Cleveland, 1931)
6. T. Ohzuku, Z. Thakehara, S. Yoshizawa, Metal-oxide of group-V-VIII as cathode materials for non-aqueous lithium cell. Denki Kagaku **46**, 411–415 (1978)
7. B. Di Pietro, M. Patriarca, B. Scrosati, On the use of rocking chair configurations for cyclable lithium organic electrolyte batteries. J. Power Sour. **8**, 289–299 (1982)

8. K.M. Abraham, D.M. Pasquariello, E.B. Willstaedt, Rechargeable sodium batteries. J. Electrochem. Soc. **137**, 1189–1190 (1990)
9. J.C. Anderson, M. Schieber, Order-disorder transitions in heat-treated rock-salt lithium ferrite. J. Phys. Chem. Solids **25**, 961–968 (1964)
10. Y. Sakurai, H. Arai, J. Yamaki, Preparation of electrochemically active alpha-$LiFeO_2$ at low temperature. Solid State Ionics **113**, 29–34 (1998)
11. B. Fuchs, S. Kemmler-Sack, Synthesis of $LiMnO_2$ and $LiFeO_2$ in molten Li halides. Solid State Ionics **68**, 279–285 (1994)
12. T. Shirane, R. Kanno, Y. Kawamoto et al., Structure and physical properties of lithium iron oxide, $LiFeO_2$, synthesized by ionic exchange reaction. Solid State Ionics **79**, 227–233 (1995)
13. M. Tabuchi, C. Masquelier, T. Takeuchi et al., Li^+/Na^+ exchange from alpha-$NaFeO_2$ using hydrothermal reaction. Solid State Ionics **90**, 129–132 (1996)
14. K. Ado, M. Tabuki, H. Kobayashi et al., Preparation of $LiFeO_2$ with alpha-$NaFeO_2$-type structure using a mixed-alkaline hydrothermal method. J. Electrochem. Soc. **144**, L177–L180 (1997)
15. R. Kanno, T. Shirane, Y. Kawamoto et al., Synthesis, structure, and electrochemical properties of a new lithium iron oxide, $LiFeO_2$, with a corrugated layer structure. J. Electrochem. Soc. **143**, 2435–2442 (1996)
16. J. Kim, M. Manthiram, Synthesis and lithium intercalation properties of nanocrystalline lithium iron oxides. J. Electrochem. Soc. **146**, 4371–4374 (1999)
17. A. Dugast, R. Brec, G. Ouvrard et al., Li_2FeS_2, a cathodic material for lithium secondary battery. Solid State Ionics **5**, 375–378 (1981)
18. M.S. Whittingham, Chemistry of intercalation compounds: Metal guests in chalcogenide hosts. Prog. Solid State Chem. **12**, 41–99 (1978)
19. R. Brec, A. Dugast, Chemical and electrochemical study of the $LixFeS_2$ cathodic system ($0 < x < 2$). Mater. Res. Bull. **15**, 619–625 (1980)
20. K. Kanamura, C. Zhen, H. Sakaebe et al., The discharge and charge characteristics of FeOCl modified by aniline in water. J. Electrochem. Soc. **138**, 331–332 (1991)
21. H. Arai, S. Okada, Y. Sakurai et al., Cathode performance and voltage estimation of metal trihalides. J. Power Sour. **68**, 716–719 (1997)
22. A. Manthiram, J.B. Goodenough, Lithium insertion into $Fe_2(MO_4)_3$ frameworks: Comparison of M = W with M = Mo. J. Solid State Chem. **71**, 349–360 (1987)
23. A. Manthiram, J.B. Goodenough, Lithium insertion into $Fe_2(SO_4)_3$ frameworks. J. Power Sour. **26**, 403–408 (1989)
24. A.K. Padhi, K.S. Nanjundaswamy, C. Masquelier et al., Effect of structure on the Fe^{3+}/Fe^{2+} redox couple in iron phosphates. J. Electrochem. Soc. **144**, 1609–1613 (1997)
25. G.H. Li, H. Azuma, M. Tohda, $LiMnPO_4$ as the cathode for lithium batteries. Electrochem. Solid State **5**, A135–A137 (2002)
26. J. Barker, M.Y. Saidi, J.L. Swoyer, Electrochemical insertion properties of the novel lithium vanadium fluorophosphate, $LiVPO_4F$. J. Electrochem. Soc. **150**, A1394–A1398 (2003)
27. K. Amine, H. Yasuda, M. Yamachi, Olivine $LiCoPO_4$ as 4.8 V electrode material for lithium batteries. Electrochem. Solid State **3**, A178–A179 (2000)
28. J. Wolfenstine, J. Allen, $LiNiPO_4$-$LiCoPO_4$ solid solutions as cathodes. J. Power Sour. **136**, 150–153 (2004)
29. B.L. Cushing, J.B. Goodenough, $Li_2NaV_2(PO_4)_3$: A 3.7 V lithium-insertion cathode with the rhombohedral NASICON structure. J. Solid State Chem. **162**, 176–181 (2001)
30. M. Sato, H. Ohkawa, K. Yoshida et al., Enhancement of discharge capacity of $Li_3V_2(PO_4)_3$ by stabilizing the orthorhombic phase at room temperature. Solid State Ionics **135**, 137–142 (2000)
31. A.K. Padhi, K.S. Nanjundaswamy, J.B. Goodenough, Phospho-olivines as positive-electrode materials for rechargeable lithium batteries. J. Electrochem. Soc. **144**, 1188–1194 (1997)
32. O.V. Yakubovich, M.A. Simonov, N.V. Belov, The crystal structure of a synthetic triphylite $LiFe[PO_4]$. Sov. Phys. Dokl. **2**, 347–350 (1977)

33. D. Morgan, A. Van der Ven, G. Ceder, Li conductivity in LixMPO$_4$ (M = Mn, Fe, Co, Ni) olivine materials. Electrochem. Solid State **7**, A30–A32 (2004)
34. A.V. Churikov, A.V. Ivanishchev, I.A. Ivanishcheva et al., Determination of lithium diffusion coefficient in $LiFePO_4$ electrode by galvanostatic and potentiostatic intermittent titration techniques. Electrochim. Acta. **55**, 2939–2950 (2010)
35. P.P. Prosini, M. Lisi, D. Zane et al., Determination of the chemical diffusion coefficient of lithium in $LiFePO_4$. Solid State Ionics **148**, 45–51 (2002)
36. S.-Y. Chung, J.T. Bloking, Y.-M. Chiang, Electronically conductive phospho-olivines as lithium storage electrodes. Nat. Mater. **1**, 123–128 (2002)
37. J.F. Ni, H.H. Zhou, J.T. Chen et al., $LiFePO_4$ doped with ions prepared by co-precipitation method. Mater. Lett. **59**, 2361–2365 (2005)
38. G.X. Wang, S. Needham, J. Yao et al., A study on $LiFePO_4$ and its doped derivatives as cathode materials for lithium-ion batteries. J. Power Sour. **159**, 282–286 (2006)
39. S.Q. Shi, L.J. Liu, C.Y. Ouyang et al., Enhancement of electronic conductivity of $LiFePO_4$ by Cr doping and its identification by first-principles calculations. Phys. Rev. B **68**, 195108–1/5 (2003)
40. B.L. Ellis, M. Wagemaker, F.M. Mulder et al., Comment on Aliovalent Substitutions in Olivine Lithium Iron Phosphate and Impact on Structure and Properties. Adv. Funct. Mater. **20**, 186–188 (2010)
41. R. Dominkó, M. Bele, M. Gaberscek et al., Porous olivine composites synthesized by sol-gel technique. J. Power Sour. **153**, 274–280 (2006)
42. S. Yang, P.Y. Zavalij, M.S. Whittingham, Hydrothermal synthesis of lithium iron phosphate cathodes. Electrochem. Commun. **3**, 505–508 (2001)
43. J. Barker, M.Y. Saidi, J.L. Swoyer, Lithium iron(II) phospho-olivines prepared by a novel carbothermal reduction method. Electrochem. Solid State **6**, A53–A55 (2003)
44. J.X. Zhang, M.Y. Xu, X.W. Cao et al., A synthetic route for lithium iron phosphate prepared by improved coprecipitation. Funct. Mater. Lett. **3**, 177–180 (2010)
45. V. Palomares, A. Goni, I.G.D. Muro et al., New freeze-drying method for $LiFePO_4$ synthesis. J. Power Sour. **171**, 879–885 (2007)
46. C. Delmas, M. Maccario, L. Crogunnec et al., Lithium deintercalation in $LiFePO_4$ nanoparticles via a domino-cascade model. Nat. Mater. **7**, 665–671 (2008)
47. A. Manthiram, A.V. Murugan, A. Sarkar et al., Nanostructured electrode materials for electrochemical energy storage and conversion. Energ. Environ. Sci. **1**, 621–638 (2008)
48. D. Jugovic, D. Uskokovic, A review of recent developments in the synthesis procedures of lithium iron phosphate powders. J. Power Sour. **190**, 538–544 (2009)
49. Z. Li, D. Zhang, F. Yang, Developments of lithium-ion batteries and challenges of $LiFePO_4$ as one promising cathode material. J. Mater. Sci. **44**, 2435–2443 (2009)
50. A. Yamada, S.C. Chung, K. Hinokuma, Optimized $LiFePO_4$ for lithium battery cathodes. J. Electrochem. Soc. **148**, A224–A229 (2001)
51. A.S. Andersson, J.O. Thomas, B. Kalska et al., Thermal stability of $LiFePO_4$-based cathodes. Electrochem. Solid State **3**, 66–68 (2000)
52. A.S. Andersson, J.O. Thomas, The source of first-cycle capacity loss in $LiFePO_4$. J. Power Sour. **97–98**, 498–502 (2001)
53. P.P. Prosini, M. Carewska, S. Scaccia et al., A new synthetic route for preparing $LiFePO_4$ with enhanced electrochemical performance. J. Electrochem. Soc. **149**, A886–A890 (2002)
54. C. Delacourt, P. Poizot, S. Levasseur et al., Size effects on carbon-free $LiFePO_4$ powders. Electrochem. Solid State **9**, A352–A355 (2006)
55. M. Tang, W.C. Carter, J.F. Belak et al., Modeling the competing phase transition pathways in nanoscale olivine electrodes. Electrochim. Acta. **56**, 969–976 (2010)
56. N. Ravet, J.B. Goodenough, S. Besner et al., Improved iron based cathode material. In *Proceeding of 196th ECS Meeting*, Hawaii, 17–22 Oct 1999
57. P.P. Prosini, D. Zane, M. Pasquali, Improved electrochemical performance of a $LiFePO_4$-based composite cathode. Electrochim. Acta. **46**, 3517–3523 (2001)

58. H. Huang, S.-C. Yin, L.F. Nazar, Approaching theoretical capacity of $LiFePO_4$ at room temperature at high rates. Electrochem. Solid State **4**, A170–A172 (2001)
59. Z. Chen, J.R. Dahn, Reducing carbon in $LiFePO_4$/C composite electrodes to maximize specific energy, volumetric energy, and tap density. J. Electrochem. Soc. **149**, A1184–A1189 (2002)
60. X.J. Chen, G.S. Cao, X.B. Zhao et al., Electrochemical performance of $LiFe_{1-x}V_xPO_4$/carbon composites prepared by solid-state reaction. J. Alloys Compd. **463**, 385–389 (2008)
61. M.M. Doeff, J.D. Wilcox, R. Yu et al., Impact of carbon structure and morphology on the electrochemical performance of $LiFePO_4$/C composites. J. Solid State Electron. **12**, 995–1001 (2008)
62. R. Dominkó, M. Gaberscek, M. Bele et al., Carbon nanocoatings on active materials for Li-ion batteries. J. Eur. Ceram. Soc. **27**, 909–913 (2007)
63. F. Croce, A. D'Epifanio, J. Hassoun et al., A Novel concept for the synthesis of an improved $LiFePO_4$ lithium battery cathode. Electrochem. Solid State **5**, A47–A50 (2002)
64. A. Yamada, M. Hosoya, S.C. Chung et al., Olivine-type cathodes: Achievements and problems. J. Power Sour. **119**, 232–238 (2003)
65. M. Takahashi, S. Tobishima, K. Takei et al., Characterization of $LiFePO_4$ as the cathode material for rechargeable lithium batteries. J. Power Sour. **97–98**, 508–511 (2001)

Chapter 2
Triphylite

2.1 Carbon Added $LiFePO_4$

Carbon added $LiFePO_4$ was prepared by the solid-state reaction of Li_2CO_3 (99.9%, Fluka), $Fe(II)C_2O_4 \cdot H_2O$ (99%, Aldrich), and $(NH_4)_2HPO_4$ (99%, Fluka). The precursors were firstly decomposed and carbon powder was added before the formation of the crystalline phase. To prevent oxidation of the iron, the synthesis was performed under a flow of nitrogen gas. The starting materials were weighed in stoichiometric amounts and homogenized using a mixer. To decompose the oxalate and the phosphate, the mixture was placed in a tubular furnace and heated at 300°C for 20 h. The powder was cooled at room temperature and mixed with high-surface area carbon-black (Ketlen Jack Blak, Akzo Nobel, surface area 1250 $m^2\ g^{-1}$). After grinding and homogenization, the mixture was transferred to the furnace and annealed at 800°C for 16 h under nitrogen flow. After this time, the powder was allowed to cool at room temperature. The crystallographic structure was confirmed by X-ray diffraction. Elemental composition of the samples (Li, Fe, and P) was determined by flame and graphite atomic absorption spectrometry (AAS) using a Varian 220 FS. To determinate the carbon content, a know amount of sample was dissolved in hydrochloric acid (Carlo Erba, Analytic grade), the solution was filtered and the residue was washed with distilled water, dried, and weighed. The morphology of the samples was observed by scanning electron microscopy (SEM).

To prepare the electrodes for the electrochemical characterization, the composite cathode powder was mixed with Teflon (DuPont) in the weight ratio 97:3. The mixture was rolled into a thin sheet of uniform thickness from which 1.0 cm diameter pellets were cut. The electrode weight ranged from 6.9 to 20.6 mg. Total electrical conductivity measurements were carried out by the 2-points ac technique using a Frequency Response Analyzer (FRA Solartron mod. 1260). Battery cells were assembled in a T-shaped hydraulic connectors and stainless-steel cylinders were used as the current-collectors; lithium metal was used both as a counter and a reference electrode. Glass fibre disk was used as a separator. The cells were filled

DOI: 10.1007/978-0-85729-745-7_2,

Fig. 2.1 SEM micrographs of the samples obtained by adding **a** 5 wt% and **b** 10 wt% carbon-black to the starting material. Reproduced by permission of Elsevier Ref. [4]

with a 1M solution of $LiPF_6$ (Merck, battery grade) in ethylene carbonate:dimethyl carbonate (EC/DMC) 1:1. The cells were assembled in a dry-room (R.H. < 0.1% at 20°C). Charge/discharge tests were performed using a Maccor Battery Cycler. The electrochemical tests were conducted at temperature ranging from 20 up to 80°C.

2.2 Physical and Electrochemical Characterization of Carbon Added $LiFePO_4$

In 1997 Padhi et al. [1] showed that lithium can be electrochemically extracted from $LiFePO_4$ and inserted into $FePO_4$ along a flat potential at 3.5 V vs. Li. In 1999, Ravet et al. [2] proposed the use of an organic compound (sucrose) as a carbon source to prepare in situ modified carbon-coated material with increased electrochemical performance. Bruce et al. [3] prepared lithium manganese oxide by a low-temperature solution route which included the addition of a small amount of carbon to the solution. In this chapter, we present the same beneficial influence on the electrochemical performance of a cathode prepared with a $LiFePO_4$ synthesized in the presence of high-surface area carbonblack [4]. Two samples containing 5 and 10 wt% carbon were prepared. The molar ratio for Li:Fe:P was found almost 1:1:1 for both the compounds while the quantity of carbon-black left in the samples after the firing treatment was 4.0 and 9.0 wt%, respectively. These results suggest that about 1.0 wt% carbon-black was lost during the firing at 800°C. Figure 2.1 shows a micrograph of the samples obtained by adding 5 wt% (a) and 10 wt% (b) carbon-black to the starting materials. The samples were analyzed as grown, without any previous grinding process. The $LiFePO_4$ prepared with 5 wt% carbon (Fig. 2.1a) consists of spherical aggregates of about 10 μm diameter. The carbon-black is uniformly dispersed between the grains, but it does not completely cover the grains. Figure 2.1b refers to the sample obtained using 10 wt% carbon-black. The grain structure of the material is very similar to the previous one, but

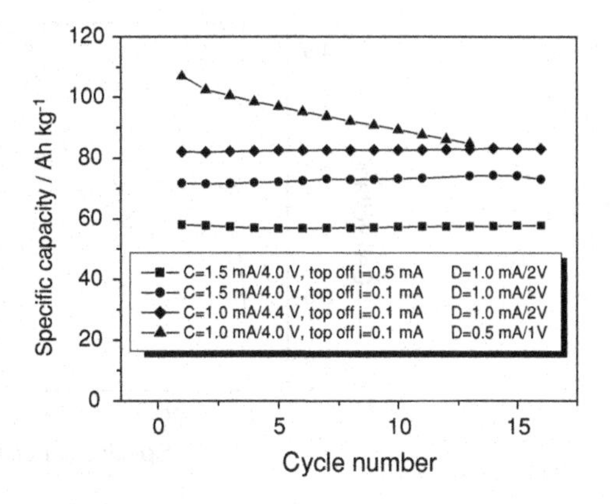

Fig. 2.2 Specific capacity versus cycle number for a $LiFePO_4$ composite cathode prepared from the sample containing 5 wt% carbon-black under different charge/discharge conditions. The cathode loading of $LiFePO_4$ was 20 mg. Temperature: 20°C. Reproduced by permission of Elsevier Ref. [4]

the grain size is smaller. The grains have an average diameter less than 10 μm. Carbon-black completely covers the grain surfaces.

Electronic conductivity of the pellets prepared with the samples obtained using 5 and 10 wt% carbon-black was very high (1.4×10^{-3} and 1.7×10^{-3} S cm^{-1}, respectively) and no further conductive additive was added in the electrode formulation.

Figure 2.2 shows the specific capacity versus cycle number for a cathode prepared with the sample synthesized using 5 wt% carbon-black.

The cell was cycled under different charge/discharge conditions. The charge procedure consisted in a galvanostatic pulse (1.5 mA, 228 A kg^{-1}) to charge the cell up to a fixed voltage (usually 4.0 V). After that the cell was kept at this voltage until the current was lower than a fixed value (top-off procedure). The discharged capacity was strongly related to the charge procedure. The capacity was observed to increase from 58 to 72 Ah kg^{-1} by reducing (during the top-off process) the end-charge current from 0.5 mA (76 A kg^{-1}) to 0.1 mA (15 A kg^{-1}). Further increase in cell capacity (up to 82 Ah kg^{-1}) was obtained by reducing the galvanostatic pulse to 1 mA (152 A kg^{-1}) and increasing the top-off potential to 4.4 V. Finally, a higher capacity was achieved decreasing the discharge current to 0.5 mA (76 A kg^{-1}) and lowering the end-discharge voltage to 1.0 V versus Li. The capacity of the material reached 110 Ah kg^{-1} but a severe capacity fading reduced the capacity over few cycles at about 80 Ah kg^{-1}. This behavior can be related to the irreversible reduction of Fe^{2+} to lower valence states. From these results, the best charge procedure was selected. Cells were charged at 152 A kg^{-1}, with a top-off at 4.4 V. This voltage was applied to the cells until the current was decrease to 1/10th of its initial value (15 A kg^{-1}). Figure 2.3 shows the voltage profile for a cathode prepared with the sample synthesized using 10 wt% carbon-black. The cell was charged using the test procedure and discharged at 15 A kg^{-1} at room temperature. The cell voltage quickly increased from the end-charge potential (2.0 V) to about 3.5 V. During the galvanosatic pulse the cathode was able to accommodate about 81% of the total capacity while a further 19% was accommodated during the constant voltage step. During the discharge the voltage quickly dropped down to

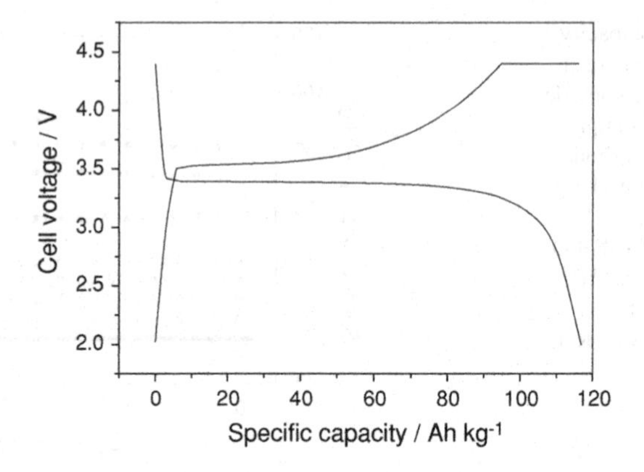

Fig. 2.3 Voltage profile for a $LiFePO_4$ composite cathode prepared from the sample containing 10 wt% carbon-black. The charge current was 152 A kg^{-1}, with a top-off at 4.4 V. This voltage was applied to the cells until the current decreased to 15 A kg^{-1}. The discharge current was 15 A kg^{-1}. The cathode loading of $LiFePO_4$ was 6.7 mg. Temperature: 20°C. Reproduced by permission of Elsevier Ref. [4]

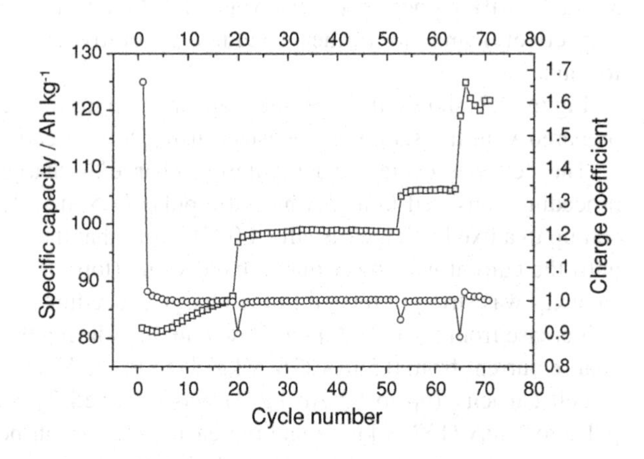

Fig. 2.4 Specific capacity (*squares*) and charge coefficient (*circles*) versus cycle number for a $LiFePO_4$ composite cathode prepared from the sample containing 10 wt% carbon-black. The cathode loading of $LiFePO_4$ was 6.7 mg. Temperature: 20°C. Reproduced by permission of Elsevier Ref. [4]

3.3 V; after which it was almost independent from the degree of reduction. When the cell was discharged at about 85% of the capacity, the voltage dropped below 3.0 V and then fell sharply. To test the effect of different discharge rates, the cells were galvanostatically discharged at different current densities corresponding to 15, 30, 60 and 152 A kg^{-1} at various temperatures, ranging from 20 up to 80°C.

Figure 2.4 shows the specific capacity and the charge coefficient plotted against cycle number. The electrochemical test was conducted at room temperature. As pointed out by Andersson et al. [5], the capacity during the first lithium extraction was higher than the capacity recovered during the following discharge cycle. The capacity loss was about 50 Ah kg^{-1} and the charge coefficient assumed a value as high as 1.66. After the first cycle, the charge coefficient tended toward unity; after the 5th cycle its value was 1.003 and this value was found constant during subsequent cycles. At the higher current density (152 A kg^{-1}) the active

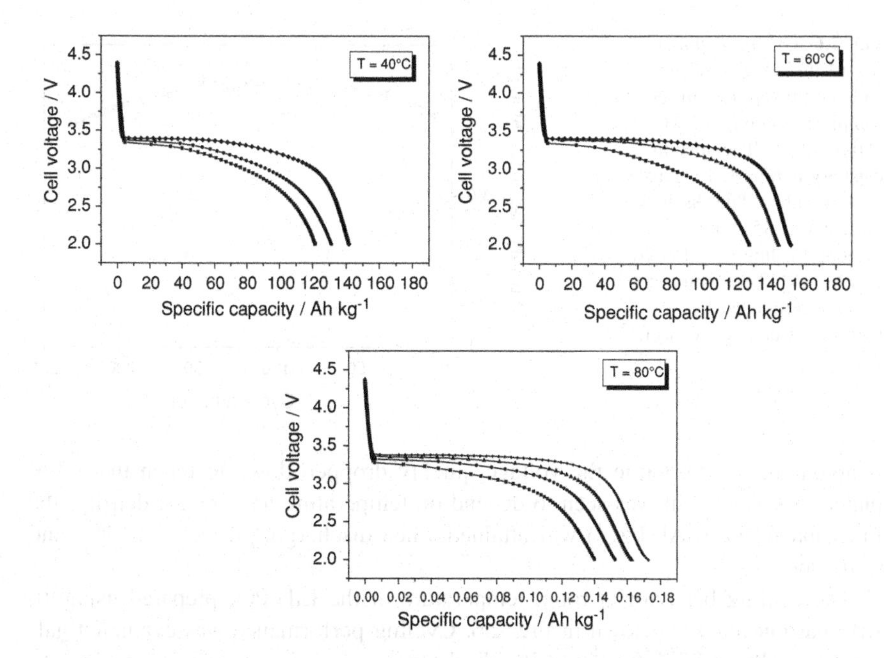

Fig. 2.5 Voltage profile for a $LiFePO_4$ composite cathode prepared using 10 wt% carbon-black at various temperatures. The charge conditions are the same as in Fig. 2.3. The discharge currents were: 15 A kg^{-1} (*diamond*), 30 A kg^{-1} (*triangle*), 60 A kg^{-1} (*circle*) and 152 A kg^{-1} (*square*). The cathode loading of $LiFePO_4$ was 6.7 mg. Reproduced by permission of Elsevier Ref. [4]

material utilization was about 50%; this value increases on decreasing the discharge current. Every time the discharge current was changed, a step was observed in the specific capacity curve. After the current variation, the charge coefficient returned to its original value. The increase of capacity produced a negative peak in the charge coefficient curve.

During the following charge, the lithium inserted in the previous step was extracted and the charge coefficient assumed a unitary value. On reducing the discharge current, the amount of lithium that can be re-inserted in the material increased, as confirmed by the negative peak in the charge coefficient and, at the lower current density (15 A kg^{-1}), the specific capacity reached the quite reasonable value of 125 Ah kg^{-1} (73% of the theoretical one and about 92% of the capacity charged during the first cycle). This result compared with the first experiences of Padhi et al. [1] is quite impressive, if we consider that the same active material utilization (73%) is reached using a current density about 8 times larger. From this behavior, it is also possible to deduce that the lithium ions extracted from the material during the first charge are only partially re-intercalated. Due to limiting interface diffusion, part of the material remains de-lithiated.

Figure 2.5 shows the voltage profiles recorded at different temperatures. The capacity was seen to increase on rising the temperature. The general appearance of the voltage profiles is similar to the previous discussed profile, recorded at room

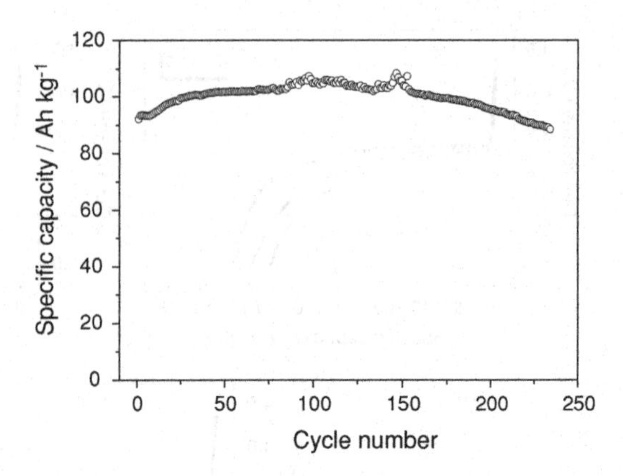

Fig. 2.6 Cycling behavior for a $LiFePO_4$ composite cathode prepared from the sample containing 10 wt% carbon-black. The specific capacity is reported versus cycle number. The discharge current was 85 A kg^{-1}. The cathode loading of $LiFePO_4$ was 6.7 mg. Temperature: 20°C. Reproduced by permission of Elsevier Ref. [4]

temperature. In discharge the voltage quickly dropped down to reach the 3.3 V plateau whose extent was seen to depend on temperature and current density: the full capacity (170 Ah kg^{-1}) was attained when discharging the cell at 80°C and C/10 rate.

The cycling behavior at room temperature of the $LiFePO_4$ prepared using 10 wt% carbon-black is shown in Fig. 2.6. Cycling performance was evaluated galvanostatically at 85 A kg^{-1} specific discharge.

The cell was galvanostatically charged using the previously reported procedure. As previously noted by Padhi et al., the specific capacity of the cell slowly increased with cycle number, reaching a maximum after 120 cycles. Thereafter, capacity fading was observed and after 230 cycles the material was able to deliver about the same capacity as during its first cycle (ca. 86 Ah kg^{-1}).

2.3 Conclusions

The addition of fine particles of carbon-black during the synthesis of $LiFePO_4$ improves the electrochemical performance of the material in terms of practical capacity and charge/discharge rate. Carbon particles, uniformly distributed between the starting materials, can interfere with the grains coalescence, decreasing the material grain size. Furthermore, the carbon increases the electric contact between the grains because the conductive filler interacts with the grains during their formation. The smaller the grain size and the grain boundary resistance, the higher the ability to support higher current densities. Carbon added $LiFePO_4$ exhibits the same active material utilization of pristine $LiFePO_4$ when discharged with current density about 8 times larger. Nevertheless, the full capacity was obtained by only cycling the material at 80°C and 15 Ah kg^{-1}. Further improvement is necessary to obtain a high-capacity material able to work at room temperature.

References

1. A.K. Padhi, K.S. Nanjundaswamy, J.B. Goodenough, Phospho-olivines as positive-electrode materials for rechargeable lithium batteries. J. Electrochem. Soc. **144**, 1188–1194 (1997)
2. N. Ravet, J.B. Goodenough, S. Besner, et al., Improved iron based cathode material. In *Proceeding of 196th ECS Meeting*, Hawaii, 17–22 Oct 1999
3. P.G. Bruce, A.R. Armstrong, H.T. Huang, New and optimised lithium manganese oxide cathodes for rechargeable lithium batteries. J. Power Sour. **68**, 19–23 (1997)
4. P.P. Prosini, D. Zane, M. Pasquali, Improved electrochemical performance of a $LiFePO_4$-based composite cathode. Electrochim. Acta. **46**, 3517–3523 (2001)
5. A.S. Andersson, J.O. Thomas, B. Kalska et al., Thermal stability of $LiFePO_4$-based cathodes. Electrochem. Solid State **3**, 66–68 (2000)

Chapter 3
Determination of the Diffusion Coefficient of $LiFePO_4$

3.1 Theory of Lithium Intercalation on $LiFePO_4$

The intercalation/de-intercalation of lithium in materials with strong electron–ion interactions proceeds following one or several reaction fronts, and leads to the coexistence of two phases [1–2]. In absence of strong electron–ion intercalation, the intercalation of lithium was described and treated similarly to an adsorption process at the metal/solution interface [3] or to the charging of electronically conductive polymers [4]. In the case of strong interactions between the intercalated species, a Frumkin-type sorption isotherm was used to describe the intercalation process and derive fundamental thermodynamic properties [5–8]. The similarities between Li intercalation and underpotential deposition (udp) behavior of various systems were discussed by Conway [2]. As for an udp process, an expression for the chemical potential of the intercalated species μ_i, as a function of the three-dimensional site occupancy fraction, X was derived. In the simplest case, for random occupancy of the lattice with no interaction, the μ_i is given by:

$$\mu_i = \mu_i^\circ + RT \ln X/(1 - X) \tag{3.1}$$

This relation is based on a model of immobile adsorption of the intercalate, i.e. Langmuir-type sorption isotherm.

A more general equation can be written, taking into account lateral interactions through a parameter g, according to a Frumkin-type sorption isotherm:

$$\mu_i = \mu_i^\circ + RT(gX + \ln X/(1 - X)) \tag{3.2}$$

Finite positive g relates to repulsive interaction; $g = 0$ signifies the absence of any interaction and Eq. 3.2 reduces to Eq. 3.1; negative values of $g > -4$ correspond to attractive interaction between the intercalation sites. At $g = -4$ a critical state arises and for $g < -4$ the interactions are so intensive that they lead to the coexistence of two-phase. The corresponding Nerst expression is:

$$E = E^\circ + RT/F\,(gX + \ln X/(1 - X)) \tag{3.3}$$

P. P. Prosini, *Iron Phosphate Materials as Cathodes for Lithium Batteries*,
DOI: 10.1007/978-0-85729-745-7_3, © Springer-Verlag London Limited 2011

The chemical diffusion coefficient under the consideration that the intercalation process follows a simple Frumkin-type sorption isotherm can be written in the form [4, 5]:

$$D = (a^2k^*)\,(1 - X)\,X\,(\partial \mu_i / \partial X)\,(kT)^{-1} \tag{3.4}$$

where (a^2k^*) represents the ionic mobility in the pure phase ($X = 1$) in terms of hopping rate constant k^*, and the nearest neighbor separation a, μ_i is the chemical potential of the intercalated species while k and T stand for the Boltzmann constant and absolute temperature, respectively.

By solving the derivative in Eq. 3.4 and by dividing the chemical diffusion coefficient for the ionic mobility, a dimensionless chemical diffusion coefficient D_{dim}, can be obtained:

$$D_{dim} = D/\,a^2k^*L = 1 + gX(1 - X) \tag{3.5}$$

where L is the Avogadro constant. According to Eq. 3.5, the model based on the Frumkin-type isotherm predicts an increase in the diffusion coefficient for repulsive interactions ($g > 0$) and a decrease for attractive interactions ($0 > g > -4$). For non-interactive systems ($g = 0$) a constant value of D is expected while for g values less than 4 a negative meaningless D value is calculated in proximity of $X = 0.5$.

The dimensionless differential capacity curve can be represented in the following form:

$$\begin{aligned} C_{dim} &= RT/F\,(\partial X / \partial (E - E^\circ)) = RT/F\,[\partial (E - E^\circ) / \partial X]^{-1} \\ &= [g + 1/X\,(1 - X)]^{-1} \end{aligned} \tag{3.6}$$

The increase in the attractive interactions results in a considerable decrease of the half-peak width of the differential capacity curve while for $g = -4$ it approaches the so called δ-function.

3.2 Experimental

Material and composite cathode preparation was described in Chap. 2. AC impedance was employed to characterize the composite cathode, using a frequency response analyzer (FRA Solartron mod. 1260). Battery cells were assembled in a T-shaped hydraulic connectors, lithium metal was used as counter electrode. A second $LiFePO_4$ composite electrode was used as reference. Glass fiber disks were used as separators, and stainless-steel cylinders as current-collectors. The cells were filled with a 1 M solution of $LiPF_6$ (Merck, battery grade) in ethylene carbonate:dimethyl carbonate (EC/DMC) 1:1. The cells were assembled in the dry room (R.H. < 0.1% at 20°C). The specific surface area was measured according to the Brunauer, Emmet, and Teller (BET) method by using a Nova 2000-Quanta Chrome apparatus.

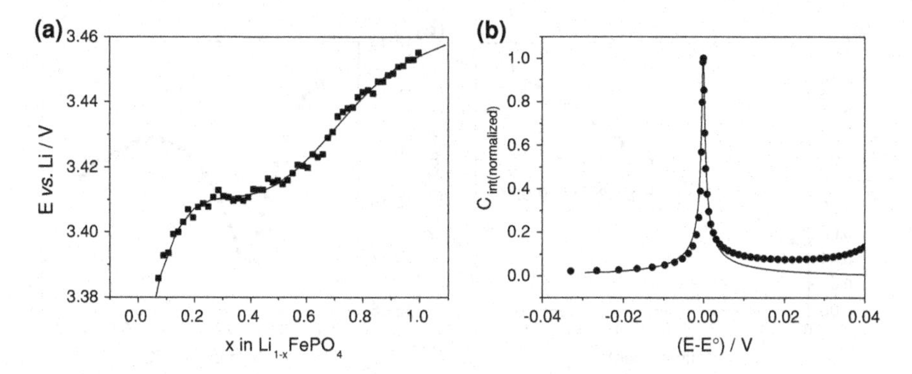

Fig. 3.1 **a** The quasi-equilibrium potential versus Li in $Li_{1-x}FePO_4$ as a function of the stoichiometry x. The quasi-equilibrium potential curve (*dot*) was fitted with a polynomial function of fifth order (*solid line*). The cathode loading of $LiFePO_4$ was 8.8 mg cm^{-2}. Temperature: 20°C. **b** A plot of the experimental differential capacity (*dot*) and the corresponding theoretical values (*solid line*) calculated according Eq. 3.6. Reproduced by permission of Elsevier Ref. [9]

3.3 Determination of the Lithium Diffusion Coefficient

The specific surface area of 10 wt% carbon added $LiFePO_4$ was surprisingly high (32.7 m^2 g^{-1}). This result can be explained considering that the carbon-black added during the synthesis totally contributes to the specific surface area of the final material. To evaluate the specific surface area we considered the sample formed by spherical aggregates of about 8 μm diameter (see Fig. 2.1b). In such a case a specific surface area of 0.208 m^2 g^{-1} was calculated. The evolution of the quasi-equilibrium potential versus x in $Li_{1-x}FePO_4$ is reported in Fig. 3.1a. From an initial value of about 3.385 V, the quasi-equilibrium potential increased monotonously upon Li ions extraction to reach a plateau in the intercalation range $0.3 < x < 0.5$ at about 3.412 V. After that the slope of the curve versus x was found constant in the Li composition range up to 1.0. The curve was fitted with a polynomial function of fifth order. The corresponding differential capacity is plotted in Fig. 3.1b together with the calculated values according to Eq. 3.6. The best fit was obtained by using $g = -3.8$. The g value is very close to the critical one ($g_{crit} = -4$) revealing that the intercalation of lithium in $LiFePO_4$ is characterized by strong interactions between the intercalated ions and the sites of the intercalation material. These strong interactions can lead to a first-order phase transition, with the occurrence of two crystallographic different phases as evidenced from the in situ XRD data [1]. Under quasi-equilibrium conditions, at the end of the relaxation step, we expect to obtain two co-existing phases at each $LiFePO_4$ particle, with relative distinct boundaries between them. The question arises to what extent the semi-infinite diffusion model is valid for situations when the intercalation proceeds partially via a movement of the inter-phase boundaries was discussed by McKinnon and Haering [4].

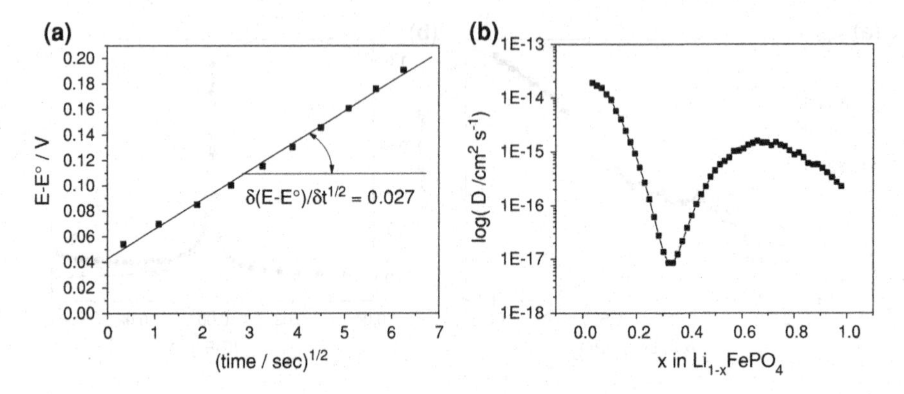

Fig. 3.2 **a** Representation of the transient voltage of the galvanostatic pulse as a function of the square root of the time for $Li_{0.6}FePO_4$. The slope of the curve was found to be 0.027 V $sec^{1/2}$. **b** The plot of the lithium chemical diffusion coefficients obtained by GITT as a function of lithium content x in $Li_{(1-x)}FePO_4$. The experimental conditions were the same as those of Fig. 3.1 Reproduced by permission of Elsevier Ref. [9]

They found that it was not possible to distinguish between two different diffusion models based on continuous (solid-solution formation) or non-continuous (two phases formation) charging procedure. Hence, in the literature one can frequently find data on D for intercalation systems with two-phase reactions [6–11]. Anyhow, in this last case the chemical diffusion coefficient mainly reflects attractive interactions between the intercalation species in the layer in which the boundary between the co-existing phases moves.

GITT was used to evaluate the lithium intercalation–deintercalation process in $LiFePO_4$. For this experiment we used current pulses of 0.25 mA during 300 s (each one corresponding to x = 0.0178 in Li_xFePO_4) followed by a potential relaxation step at open circuit until the cell voltage variation was less than 4 mV h^{-1}. The chemical diffusion coefficient of lithium D_{Li} was calculated according to Eq. 3.7 derived by Weppner and Huggins [12]:

$$D_{Li} = 4/\pi\,(V_M/SF)^2\left[I°(\delta E/\delta x)/\left(\delta E/\delta t^{1/2}\right)\right]^2 \quad \text{for } t << \tau \tag{3.7}$$

where V_M is the phosphate molar volume (44.11 cm^3 mol^{-1}), S is the contact area between electrolyte and sample (14.38 cm^2), F is the Faraday constant (96486 Coulomb mol^{-1}), I° is the applied constant electric current (2.5×10^{-4}A), $\delta E/\delta x$ is the slope of the coulometric titration curve while $\delta E/\delta t^{1/2}$ is the slope of the short-time transient voltage change. The equation is valid for times shorter than the diffusion time $\tau=(d/2\pi)^2/D$ where d is the average diameter of the grains.

Figure 3.2a reports an example of the E−E° versus $t^{1/2}$ plot recorded for $Li_{0.6}FePO_4$ after application of 0.25 mA pulse.

The plot was found linear during the first 40 s with a slope of 0.027 V $sec^{-1/2}$. Figure 3.2b shows the chemical diffusion coefficients as a function of x in $Li_{1-x}FePO_4$ obtained by substitution of the $\delta E/\delta t^{1/2}$ slopes, calculated for different

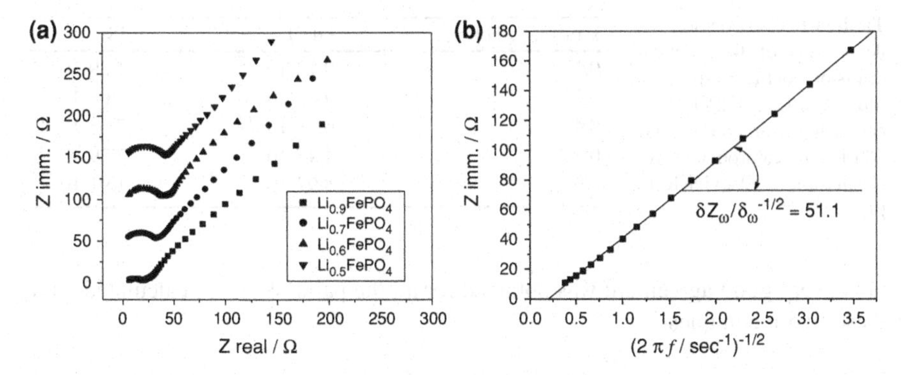

Fig. 3.3 **a** Impedance spectra for $Li_{1-x}FePO_4$ at various lithium content x. Frequency range: 0.01 Hz–10 kHz. **b** The plot of the imaginary resistance as a function of the inverse square root of angular speed for $Li_{0.6}FePO_4$. Data obtained from Impedance spectroscopy. Reproduced by permission of Elsevier Ref. [9]

values of x in Eq. 3.7. The diffusion coefficient was found to range from 1.8×10^{-14} to 2.2×10^{-16} cm^2sec^{-1} for $LiFePO_4$ and $FePO_4$, respectively with a minimum in correspondence of the peak of the differential capacity. The minimum in the chemical diffusion coefficient was predicted by the model for strong attractive interactions between the intercalation species and the host matrix. The chemical diffusion coefficient values are relatively low when compared to the diffusion coefficient of other active materials used in lithium battery [6–11, 13–15]. Figure 3.3a shows Cole–Cole plots under open-circuit condition for four different compositions in $Li_{1-x}FePO_4$. The semicircles have a high-frequency intercept that identifies the ionic conductivity of the electrolyte. At lower frequencies, the resistance related to the charge transfer between the electrolyte and the active material can be identified. At very low frequencies, there is a third region in which a typical Warburg behavior, related to the diffusion of lithium ions in the cathode active material, is seen. By using the model proposed by Ho et al. [16] the diffusion coefficient for $Li_{1-x}FePO_4$ was calculated by using Eq. 3.8:

$$D_{Li} = 1/2[(V_M/SFA)(\delta E/\delta x)]^2 \qquad (3.8)$$

where V_M is the phosphate molar volume (44.11 cm^3 mol^{-1}), S is the contact area between electrolyte and sample (14.38 cm^2), F is the Faraday constant (96486 Coulomb mol^{-1}), $\delta E/\delta x$ is the slope of the coulometric titration curve while A was obtained from the Warburg impedance.

Figure 3.3b shows the plot of the imaginary resistance determined by IS as a function of the inverse square root of the angular frequency for $Li_{0.6}FePO_4$. A linear behavior was observed for frequency values ranging from 2.5 Hz to 13 mHz with a slope of 51.1Ω $sec^{1/2}$. The diffusion coefficient of lithium for different x values in $Li_{1-x}FePO_4$ obtained by substitution of the curve slopes in Eq. 3.8 is reported in Table 3.1. In the same table the corresponding values obtained by GITT calculation are reported for comparison. As demonstrated in this

Table 3.1 Composition dependence of the chemical diffusion coefficient of lithium in $Li_{1-x}FePO_4$ calculated from IS data and GITT data. Reproduced by permission of Elsevier Ref. [9]

x in $Li_{(1-x)}FePO_4$	GITT	IS
0.1	$9.13\ 10^{-15}$	$1.29\ 10^{-14}$
0.2	$9.32\ 10^{-16}$	$1.08\ 10^{-15}$
0.4	$6.48\ 10^{-17}$	$7.68\ 10^{-17}$
0.5	$4.47\ 10^{-16}$	$7.39\ 10^{-16}$
0.9	$4.97\ 10^{-16}$	$1.91\ 10^{-15}$

table, very good agreement was obtained for all the pairs of values calculated with these two techniques.

3.4 Conclusions

GITT and IS were used to determinate the diffusion coefficient of lithium in $LiFePO_4$ as a function of the lithium content. Although the theory of GITT and IS was proven to be strictly valid for solid-solution reactions, reasonable effective values of D can also be obtained in the case of two-phase reactions if the interactions among the intercalation sites are moderate. The lithium reaction in the material was modelled in terms of a Frumkin-type sorption isotherm, taking into account of host–guest interactions through a parameter g. By fitting the experimental differential capacity with the theoretical one, we estimated an interaction parameter $g = -3.8$ that is less than the critical one ($g_{crit.} = -4$). The calculated D_{Li} as a function of x in $Li_{1-x}FePO_4$ was found to range from 1.8×10^{-14} to 2.2×10^{-16} $cm^2\ s^{-1}$ for $LiFePO_4$ and $FePO_4$, respectively with a minimum in correspondence of the peak of the differential capacity. The D_{Li} obtained by IS agreed very well within the same order of magnitude. The relatively low value of the calculated diffusion coefficient allows us to state that slow lithium-ion diffusion in $LiFePO_4$ is the main cause of the poor electrochemical performance exhibited from the material. The reduction of the grain size could be one of the possible routes to enhance the performance of $LiFePO_4$ to make it feasible as a cathode of high-power density lithium–ion batteries.

References

1. A.S. Andersson, J.O. Thomas, B. Kalska et al., Thermal stability of $LiFePO_4$-based cathodes. Electrochem. Solid State Lett. **3**, 66–68 (2000)
2. B.E. Conway, Two-dimensional and quasi-two-dimensional isotherms for Li intercalation and upd processes at surfaces. Electrochim. Acta. **38**, 1249–1258 (1993)
3. M.A. Vorotyntsev, J.P. Badiali, Short-range electron–ion interaction effects in charging the electroactive polymer films. Electrochim. Acta. **39**, 289–306 (1994)
4. W.R. McKinnon, R.R. Haering, *Modern Aspect in Electrochemistry*, vol. 15 (Plenum Press, New York, 1987)

5. M.D. Levi, G. Salitra, B. Markovsky et al., Solid-state electrochemical kinetics of Li-ion intercalation into $Li_{1-x}CoO_2$: Simultaneous application of electroanalytical techniques SSCV, PITT, and EIS. J. Electrochem. Soc. **146**, 1279–1289 (1999)
6. Y. Sato, T. Asada, H. Tokugawa et al., Observation of structure change due to discharge/charge process of V_2O_5 prepared by ozone oxidation method, using in situ X-ray diffraction technique. J Power Sour. **68**, 674–679 (1997)
7. M. Nishizawa, R. Hashitani, T. Itoh et al., Measurements of chemical diffusion coefficient of lithium ion in graphitized mesocarbon microbeads using a microelectrode. Electrochem. Solid State **1**, 10–12 (1998)
8. J. Barker, R. Pynenburg, R. Koksbang, Determination of thermodynamic, kinetic and interfacial properties for the Li//$Li_xMn_2O_4$ system by electrochemical techniques. J. Power Sour. **52**, 185–192 (1994)
9. P.P. Prosini, M. Lisi, D. Zane et al., Determination of the chemical diffusion coefficient of lithium in $LiFePO_4$. Solid State Ionics **148**, 45–51 (2002)
10. A.V. Churikov, A.V. Ivanishchev, I.A. Ivanishcheva et al., Determination of lithium diffusion coefficient in $LiFePO_4$ electrode by galvanostatic and potentiostatic intermittent titration techniques. Electrochim. Acta. **55**, 2939–2950 (2010)
11. S.-I. Pyun, J.-S. Bae, The ac impedance study of electrochemical lithium intercalation into porous vanadium oxide electrode. Electrochim. Acta. **41**, 919–925 (1996)
12. W. Weppner, R.A. Huggins, Determination of the kinetic parameters of mixed-conducting electrodes and application to the system Li_3Sb. J. Electrochem. Soc. **124**, 1569–1578 (1977)
13. L. Li, G. Pistoia, Secondary Li cells. II. Characteristics of lithiated manganese oxides synthesized from $LiNO_3$ and MnO_2. Solid State Ionics **47**, 241–249 (1991)
14. F. Coustier, S. Passerini, W.H. Smyrl, Dip-coated silver-doped V_2O_5 xerogels as host materials for lithium intercalation. Solid State Ionics **100**, 247–258 (1997)
15. M.D. Levi, K. Gamolsky, D. Aurbach et al., Determination of the Li ion chemical diffusion coefficient for the topotactic solid-state reactions occurring via a two-phase or single-phase solid solution pathway. J. Electroanal. Chem. **477**, 32–40 (1999)
16. C. Ho, I.D. Raistrick, R.A. Huggins, Application of A-C techniques to the study of lithium diffusion in tungsten trioxide thin films. J. Electrochem. Soc. **127**, 343–350 (1980)

Chapter 4
Vivianite and Beraunite

4.1 Synthesis of $Fe_3(PO_4)_2 \times nH_2O$ and $3Fe2O_3 \bullet 2P_2O_5 \bullet 10H_2O$

De-ionized water (18 MΩ cm^{-1}) produced by a Milli-Q water production system (Millipore, Bedford, MA) was used to prepare all solutions. $Fe(NH_4)_2(SO_4)_2 \cdot 6H_2O$ (Carlo Erba, RPE) and K_2HPO_4 (Carlo Erba, RPE) crystalline solids were dissolved in deionized water to make iron and phosphate stock solutions. Stock standard solutions of iron (Aldrich) and phosphorus (Aldrich) at 1000 ppm were used to prepare the working standard solutions for the AAS determination of the iron and phosphorus. The stock solutions were standardized by using AAS.

To prepare the iron(II)phosphate a solution of 0.06 M $Fe(NH_4)_2(SO_4)_2 \cdot 6H_2O$ (Carlo Erba, RPE) was added at ambient temperature to a constantly stirred 0.04 M solution of K_2HPO_4 (Carlo Erba, RPE), in a 1:1 volume proportion. A pale-blue gel started to form after the addition was completed. The gel was collected on a membrane filter (0.8 μm), washed several times with de-ionized water, and dried in air in an oven at 100°C. After the heating treatment, the color of the powder changed from pale-blue to dark-yellow. Hereinafter, we shall refer to this material as "as prepared".

A simultaneous TG-DTA apparatus SDT 2960 (TA Instruments) was used for thermal characterization. Samples ranging between 5 and 10 mg in weight were heated over a temperature that ranged from ambient to 800°C at a heating rate of 5°C min^{-1} in air atmosphere at 100 ml min^{-1} flow rate. α-Al_2O_3 was used as a reference material. Samples were run in open platinum pans.

The elemental composition of the precipitates (Fe, P) was determined by means of flame and graphite AAS (Varian 220 FS).

XRD spectra of the powder were carried out by an X' PERT-MPD diffractometer using Cu-kα radiation. The patterns were acquired by a theta-2theta goniometer mounted on the line shape radiation and equipped by a monochromator, a programmable receiving slit and a Xe-filled proportional detector. The powder was mounted in a sample stage located in a High Temperature Camber which allows carrying on measurements in the temperature range from room

P. P. Prosini, *Iron Phosphate Materials as Cathodes for Lithium Batteries*,
DOI: 10.1007/978-0-85729-745-7_4,

temperature to 1200°C in air atmosphere. To check the temperature at which the phase transition occurs, the XRD spectra of the compound were recorded at different temperatures.

Mössbauer spectra were collected both on a sample of the material as prepared and on a sample of the material after cell discharge. The samples had the shape of a 20 mm disc and an effective density of about 5 mg cm^{-2} of natural iron. A 25 mCi source of ^{57}Co in Rh matrix at room temperature was used. For each sample, spectra were collected at room temperature and at –260°C. A Gifford-MacMahon cryo-generator was used to cool the samples.

Composite cathode tapes were made by roll milling a mixture of 86% active material with 4% of binder (Teflon, DuPont) and 10% of carbon (SuperP, MMM Carbon). Electrodes were punched in the form of discs, with a diameter of 8 mm. The electrode weight ranged from 5 to 11 mg, which corresponds to an active material mass loading of 8–19 mg cm^{-2}. The electrodes were assembled in a sealed cell formed by a polypropylene T-type pipe connector with three cylindrical stainless steel (SS316) current collectors. A lithium foil was used both as an anode and a reference electrode and a glass fibre was used as a separator. The cell was filled with ethylene carbonate/diethyl carbonate 1:1 $LiPF_6$ 1M electrolyte solution. The cycling tests were carried out automatically by means of a battery cycler (Maccor 4000). Composite cathode preparation, cell assembly, test, and storage were performed in a dry-room (R.H. < 0.1% at 20°C).

4.2 Physical and Electrochemical Characterization of $3Fe_2O_3•2P_2O_5•10H_2O$

Wet chemical methods, such as the hydrothermal, template and precipitation process provide effective-cost preparation techniques, especially if compared to "high-temperature" methods [1–4]. Hydrated iron (II) phosphate was prepared through the spontaneous gelification of iron (II) and phosphate aqueous solutions at ambient temperature. The material was characterized as synthetic vivianite [$Fe_3(PO4)_2 \times nH_2O$] [5]. The material was oxidized by exposure to air at 100°C, becoming dark-yellow [6]. The chemical analysis of the oxidized material showed the presence of iron and phosphate but not the presence of alkali or sulphate ions. The results of the chemical analysis gave rise to 33.2 wt% Fe_t and 12.4 wt% P_t, suggesting the molar ratio Fe_t:P_t = 3:2.

The TG/DTG/DTA curves of the above-mentioned compound are shown in Fig. 4.1. Over the temperature range from ambient to 800°C there was a weight loss in the TG curve (Fig. 4.1(— ·)), with a corresponding peak at 115°C in the derivative curve (Fig. 4.1 (- -)).

This weight loss corresponds to the elimination of water. The mass loss was about 18 wt%. On the basis of these results it is possible to propose the following stoichiometry of the oxidized material: $3Fe_2O_3•2P_2O_5•10H_2O$. The corresponding DTA curve in Fig. 4.1(—) shows an endothermic effect around at 110°C. At a

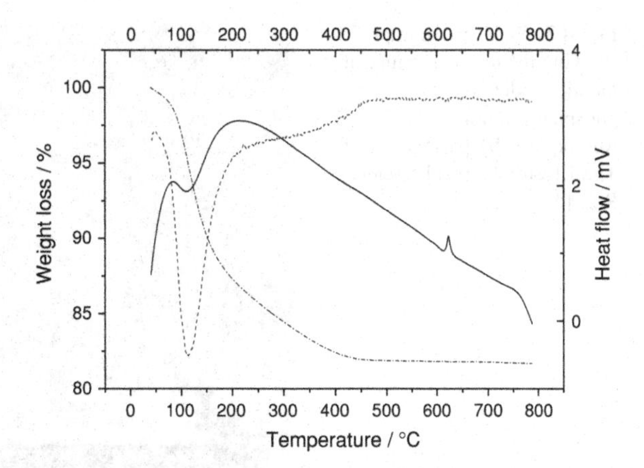

Fig. 4.1 TG (—·), DTG (--), and DTA (—) curves recorded over the temperature range from room temperature to 800°C for the oxidized sample fired at 100°C. Reproduced by permission of The Electrochemical Society Ref. [6]

higher temperature, an exothermic effect is displayed at 622°C, which is not accompanied by an appreciable weight loss in the TG curve. This effect is probably related to the crystallization of the compound.

Figure 4.2 shows the X-ray diffraction patterns of the oxidized compound before and after heated in air atmosphere, at different temperatures: 400, 500°C, and finally 650°C. The as-prepared compound is completely amorphous and keeps the same amorphous phase increasing the temperature until 500°C. On further heating at 650°C the X-ray diffraction pattern shows well defined diffraction peaks, indicating that a transformation from amorphous to crystalline occurs at this temperature. The main peaks may be attributed to the $FePO_4$ crystalline phase

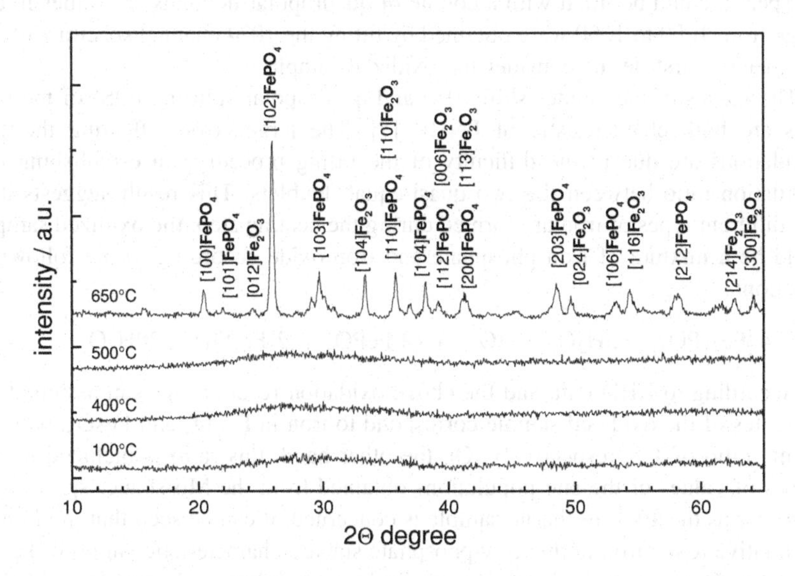

Fig. 4.2 X-ray powder diffraction patterns (Cu-Kα radiation) of the oxidized sample fired at 100, 400, 500, and 650°C. Reproduced by permission of The Electrochemical Society Ref. [6]

Fig. 4.3 SEM micrograph showing the microstructure of the air oxidized iron-phosphate sample. Reproduced by permission of The Electrochemical Society Ref. [6]

(JCPDS card n.° 29-0715). The remaining reflections can be assigned to Fe_2O_3 crystalline phase (JCPDS card n.° 33-0664).

Figure 4.3 is a SEM micrograph of the material. It is characterized by thin layers, arranged in alternating and overlapping planes, spreading out radially as a flower corolla.

The spectra obtained from the Mössbauer spectroscopy are reported in Fig. 4.4 together with their fits, which were elaborated using the RECOIL program. Since the samples are completely amorphous, magnetic components in the spectra are not expected to be found in spectrum analysis. In accordance with this, all the spectra could be fitted with a couple of quadrupolar doublets. χ^2 values in the range from 1000 to 1050 were obtained by fitting the 1024 channel spectra with 13 parameters. First, let us consider the oxidized sample.

The values of the isomer shifts (IS) and quadrupolar splitting (QS) of the two sites are both characteristic of Fe(III) [7]. The large errors affecting the site populations are due to the difficulty of the fitting procedure in establishing the population ratio between the two quasi-equal doublets. This result suggests that two different types of iron are formed during the heating step: the oxidized sample could be a mixture of iron phosphate and iron oxide according to the following equation:

$$4\left[Fe_3(PO_4)_2 \times 5H_2O\right] + 3O_2 \rightarrow (8\ FePO_4 + 2\ Fe_2O_3) \times 20H_2O \tag{4.1}$$

According to XRD data and the above oxidation reaction, we can assume that both sites of the oxidized sample correspond to iron in $FePO_4$ and Fe_2O_3, with an atomic ratio of 1:2, respectively. On the other hand, this ratio is included in the ranges of values of the site populations obtained from the Mössbauer spectra.

As far as the after discharge sample is concerned, it can be seen that the IS and QS relative to the iron in the iron–phosphate site are characteristic values of Fe(II) ions that is, practically all the iron in the iron phosphate is in the +2 oxidation state, while the iron in the iron oxide remains in the +3 oxidation state. This result

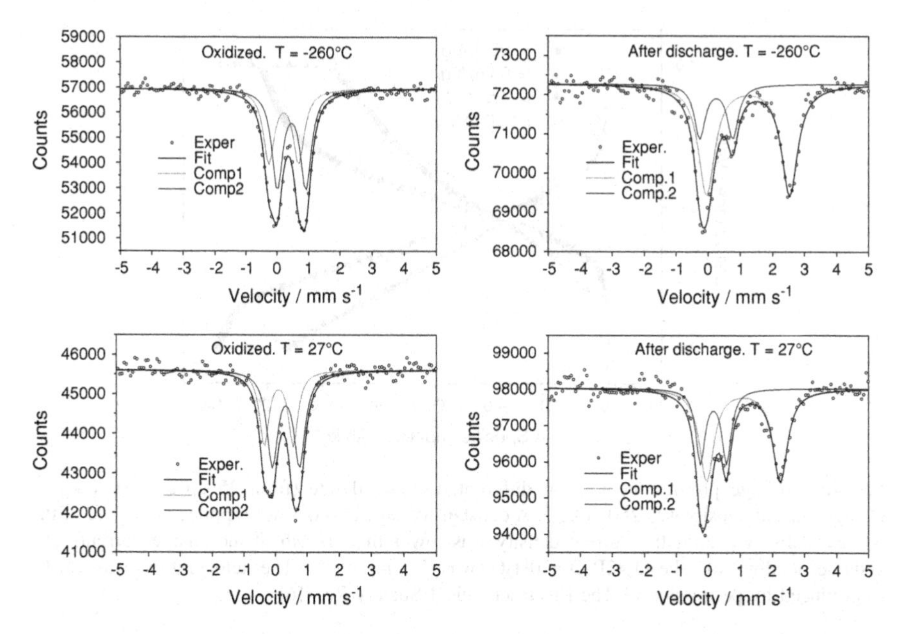

Fig. 4.4 Mössbauer spectra, for the $3Fe_2O_3 \bullet 2P_2O_5 \bullet 10H_2O$-based cathode, collected at –260°C (*top*) and at 27°C (*bottom*) before (*left*) and after (*right*) the electrochemical reduction. Reproduced by permission of The Electrochemical Society Ref. [6]

agrees with the fact that iron oxide is not electrochemically active in the potential window explored during the experiment (the end charge potential was 2.0 Volt versus Li). The voltage profiles of the material are showed in Fig. 4.5. The cell was discharged galvanostatically under different specific currents, ranged from 25 up to 250 A kg^{-1} (the corresponding values are reported in the figure). The cut-off voltage was 2.0 V. The cell was always recharged with the same procedure, to assure identical initial conditions: a constant current step (250 A kg^{-1}) until the voltage reached 4.0 V, followed by a constant voltage step until the current descended below 25 A kg^{-1}.

At the lowest discharge current density used (25 A kg^{-1}), the cell was able to deliver a specific capacity of 138 Ah kg^{-1}, based on the active material weight, corresponding to a discharge time of 5.4 h. By increasing the current density the utilization of the active material decreased, and about 105 Ah kg^{-1} was delivered in about 0.44 h at a current density 10 times higher. This result, that is not impressive when compared with other cathode active materials, appears to be very good when compared with active materials based on iron phosphate in which an increase in the discharge current results in a severe capacity fading [8]. For crystalline $LiFePO_4$, this capacity fading was claimed to have a kinetic origin because the full capacity was recovered on returning to lower current densities [9]. The excellent performance of the amorphous compound is due to the unique microstructure and the small particle size achieved by the solution-based synthesis,

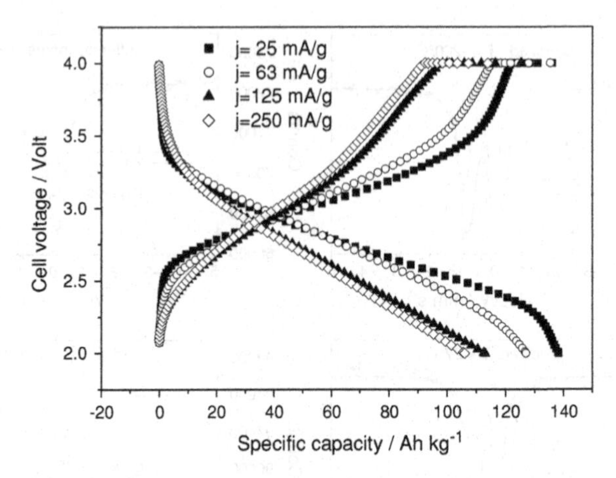

Fig. 4.5 Voltage profiles recorded at different discharged rate (from 25 up to 250 A kg^{-1}). Charge current density was 250 A kg^{-1}. A constant voltage of 4.0 V was applied at the end of the galvanostatic step until the current density was lower than 1/10th of the charge current. The cathode loading of $3Fe_2O_3{\bullet}2P_2O_5{\bullet}10H_2O$ was 7.9 mg cm^{-2}. The temperature was 20°C. Reproduced by permission of The Electrochemical Society Ref. [6]

followed by a mild oxidation reaction, compared to the high temperature procedures employed in the literature. The reduced utilization of the active material with increasing currents can be related to an increase in the ohmic drop, rather than dealing with kinetic limitation. In fact, the shoulder observed in the voltage profile for the lower current density, just before the end of the discharge, tends to reduce by increasing the current, and disappears at the highest current densities. This behavior leads us to conclude that the cell reaches the end discharge voltage before the lithium depletion on the cathode surface, due to a poor transport of the lithium from the bulk, occurs.

The reversibility of the material is illustrated in Fig. 4.6, where the specific capacity of the cathode (based on the weight of the active material) versus the cycle number is reported.

The insertion/release process was driven at 170 A kg^{-1}. Every 20 cycles, a cycle at reduced specific current (56 A kg^{-1}) was performed for testing the capacity delivered by the material under less stressed conditions. The capacities delivered by the material during the first cycles at the lower and the higher current densities were 96 and 67 Ah kg^{-1}, respectively. These values are lower than the values obtained in the previous experiment (see Fig. 4.5). The differences arise from the fact that the charge/discharge conditions were different. The voltage cut-offs were reduced from 4.0 to 3.8 Volt in charge and from 2.0 to 2.2 Volt in discharge, while the constant potential step was not imposed at the end of the charge (constant current charge/discharge condition). The lithium intercalation into the amorphous structure was seen to be very reversible. The capacity fade, evaluated by considering the capacity exhibited in the test cycles, was about 0.025% per cycle. This exceptionally low capacity fade can be related to the inherent structure of the

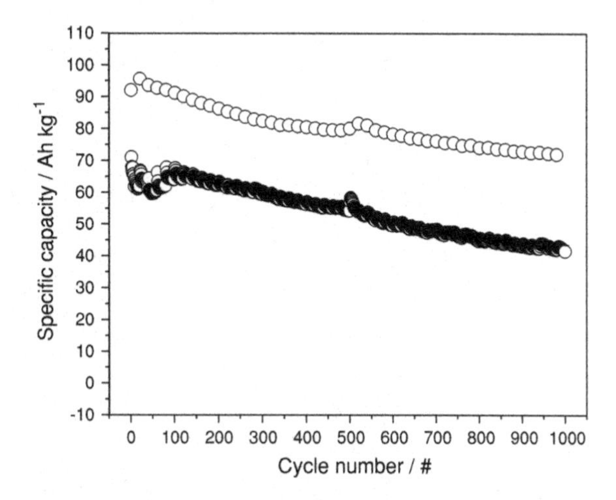

Fig. 4.6 Specific capacity upon discharge; the insertion/release process was driven at a specific current of 170 A kg^{-1} (*lower curve*). A cycle at reduced specific current (56 A kg^{-1}) was performed every 20 cycles (*upper curve*). The temperature was 20°C. The cathode loading of $3Fe_2O_3{\bullet}2P_2O_5{\bullet}10H_2O$ was 10.88 mg cm^{-2}. Reproduced by permission of The Electrochemical Society Ref. [6]

material that is retained throughout the entire intercalation range. The fatigue and the stress arising from the lithium insertion/release process are well tolerated from the material structure. In this way, the structural changes of the electrode that may cause contact loss and capacity fade upon cycling are reduced and the cycle life of the cathode is enhanced.

References

1. R. Dominkó, M. Bele, M. Gaberscek et al., Porous olivine composites synthesized by sol-gel technique. J. Power Sour. **153**, 274–280 (2006)
2. S. Yang, P.Y. Zavalij, M.S. Whittingham, Hydrothermal synthesis of lithium iron phosphate cathodes. Electrochem. Commun. **3**, 505–508 (2001)
3. J.X. Zhang, M.Y. Xu, X.W. Cao et al., A synthetic route for lithium iron phosphate prepared by improved coprecipitation. Funct. Mater Lett. **3**, 177–180 (2010)
4. V. Palomares, A. Goni, I.G.D. Muro et al., New freeze-drying method for $LiFePO_4$ synthesis. J. Power Sour. **171**, 879–885 (2007)
5. S. Scaccia, M. Carewska, A. Di Bartolomeo et al., Thermoanalytical investigation of nanocrystalline iron (II) phosphate obtained by spontaneous precipitation from aqueous solutions. Thermochim. Acta. **397**, 135–141 (2003)
6. P.P. Prosini, L. Cianchi, G. Spina et al., Synthesis and characterization of amorphous $3Fe_2O_3{\bullet}2P_2O_5{\bullet}10H_2O$ and Its electrode performance in lithium batteries. J. Electrochem. Soc. **148**, A1125–A1129 (2001)
7. N.N. Greenwood, T.C. Gibb, *Mössbauer Spectroscopy* (Chapman & Hall, London, 1975)
8. A.K. Padhi, K.S. Nanjundaswamy, C. Masquelier et al., Effect of structure on the Fe^{3+}/Fe^{2+} redox couple in iron phosphates. J. Electrochem. Soc. **144**, 1609–1613 (1997)
9. A.K. Padhi, K.S. Nanjundaswamy, J.B. Goodenough, Phospho-olivines as positive-electrode materials for rechargeable lithium batteries. J. Electrochem. Soc. **144**, 1188–1194 (1997)

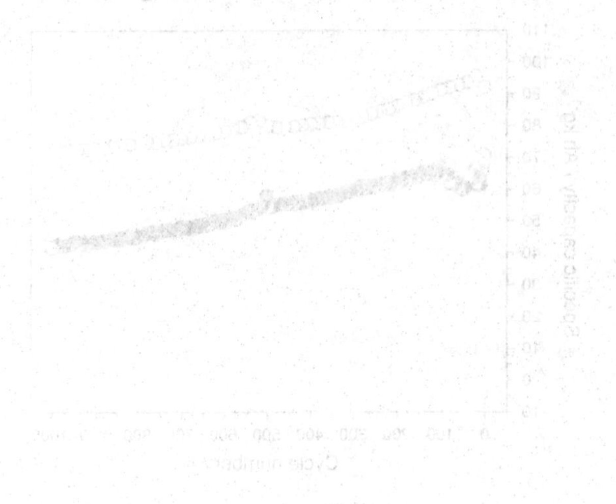

Chapter 5
Amorphous Iron Phosphate

5.1 Synthesis of Amorphous and Crystalline $FePO_4$

De-ionized water (18 MΩ cm^{-1}) produced by a Milli-Q water production system (Millipore, Bedford, MA) was used to prepare all solutions. $Fe(NH_4)_2(SO_4)_2.6H_2O$ (Carlo Erba, RPE) and $NH_4H_2PO_4$ (Carlo Erba, Reagent grade) crystalline solids were dissolved in de-ionized water. Hydrogen peroxide 30% weight (Reagent grade, Ashland Chemical Italian) was used for iron oxidation. An equimolar solution of $NH_4H_2PO_4$, in a 1:1 volume proportion was added to a solution of 0.025 M $Fe(NH_4)_2(SO_4)_2{\cdot}6H_2O$. Then 3 ml of concentrated hydrogen peroxide solution was added to the solution at ambient temperature under vigorous stirring. A white precipitate started to form immediately after the addition of hydrogen peroxide. When the precipitation was completed, the precipitate was collected on membrane filter (0.8 μm), rinsed several times with de-ionized water and dried in air in a dry-room (RH < 0.2% at 20°C) for several days.

A simultaneous TG-DTA apparatus SDT 2960 (TA Instruments) was used for thermal analysis. Samples of about 5–10 mg in weight were heated from ambient temperature to 800°C at a heating rate of 5°C min^{-1} in air atmosphere at 100 ml min^{-1} flow rate. α-Al_2O_3 was used as reference material. Samples were run in open platinum pans. Elemental composition of the precipitate (Fe, P) was determined by flame and graphite atomic absorption spectrometry (AAS) using a Varian 220 FS instrument. The as-prepared and the fired materials (in air at 400 and 650°C for 24 h) were characterized by XRD analysis (Philips PW 3710 diffractometer) using Cu-Kα radiation. The morphology of the samples was observed by SEM.

The material density was measured by using a helium picnometer (Accu pyc 1330-Micromeritics). The specific surface area was measured according to the Brunauer, Emmet and Teller (BET) method by using a Nova 2000-Quanta Chrome apparatus.

Composite cathode tapes were made by roll milling a mixture of 86 wt% active material, 4 wt% binder (Teflon, DuPont) and 10 wt% carbon (Super P,

P. P. Prosini, *Iron Phosphate Materials as Cathodes for Lithium Batteries*,
DOI: 10.1007/978-0-85729-745-7_5,

MMM Carbon). Electrodes were punched in the form of discs typically with a diameter of 10 mm. The electrode weight ranged from 5.0 to 11.0 mg corresponding to an active material mass loading of 5.5–12.0 mg cm^{-2}. The electrodes were assembled in sealed cells formed by a polypropylene *T*-type pipe connector with three cylindrical stainless steel (SS316) current collectors. A lithium foil was used both as an anode and a reference electrode and a glass fibre was used as a separator. Three-electrode cell configuration was used for GITT measurements using a lithium foil as a reference electrode. The cells were filled with ethylene carbonate/diethyl carbonate 1:1 $LiPF_6$ 1M electrolyte solution. The cycling tests were carried out automatically by means of a battery cycler (Maccor 4000). Composite cathode preparation, cell assembly, test, and storage were performed in the dry-room.

5.2 Physical and Electrochemical Characterization of Amorphous and Crystalline $FePO_4$

In the previous chapter it was shown that a mixture of amorphous $3Fe_2O_3{\bullet}2P_2O_5{\bullet}10H_2O$, prepared by air oxidation of iron(II)phosphate, exhibited good electrochemical properties as cathode of a lithium cell [1]. However, two different types of iron were formed during the heating treatment of iron(II)phosphate, namely iron(III)phosphate and iron(III)oxide. Since the latter form is electrochemically non-active for lithium intercalation in the investigated voltage range (4.0–2.0 V) [2], the electrochemical properties of such material was only due to the amorphous iron(III)phosphate phase. For this reason we investigated a new synthetic route to prepare pure amorphous iron(III)phosphate [3]. Amorphous iron(III)phosphate was synthesized by spontaneous precipitation from equimolar aqueous solutions of $Fe(NH_4)_2(SO_4)_2.6H_2O$ and $NH_4H_2PO_4$, using hydrogen peroxide as an oxidizing agent as described in Eq. 5.1:

$$\begin{aligned} &Fe(NH_4)_2(SO_4)_2 + NH_4H_2PO_4 + 0.5\ H_2O_2 \\ &\rightarrow FePO_4 + NH_3 + 2(NH_4)HSO_4 + H_2O \end{aligned} \tag{5.1}$$

The chemical analysis of the precipitate gives rise to a molar ratio Fe_t:P_t = 1:1, suggesting the general formula $FePO_4 \times H_2O$. The TG/DTA curves of the above-mentioned precipitate are displayed in Fig. 5.1. Over the temperature range from ambient to 550°C there is a weight loss in the TG curve (Fig. 5.1(—)) that corresponds to the elimination of crystalline water to give the anhydrous salt. The mass loss of 15% corresponds to the presence of 1.5 molecules of water per mole of compound. At higher temperature two exothermic peaks are displayed, namely, at 643 and 678°C, which are not accompanied by appreciable weight loss in the TG curve. These peaks likely indicate two-step structural transformation of the $FePO_4$ framework [4]. Finally, a weak endothermic effect occurs at 716°C without appreciable weight loss, which can be ascribed to the $\alpha \rightarrow \beta$ transition as reported

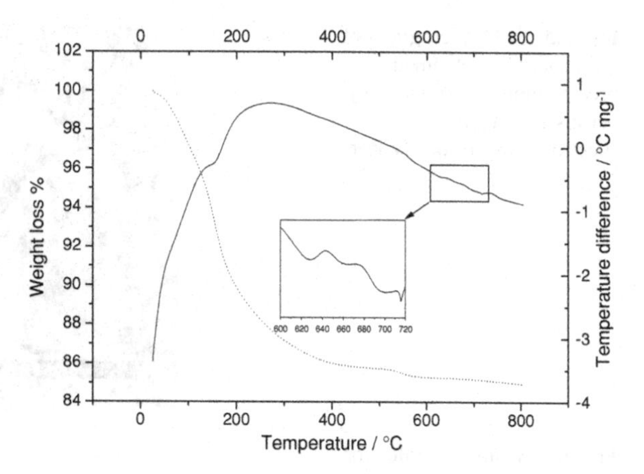

Fig. 5.1 TG (….), and DTA (—) *curves* for precipitated $FePO_4$ recorded over the temperature range from ambient to 800°C at a heating rate of 5°C min^{-1} in air (100 ml min^{-1} flow rate). Reproduced by permission of The Electrochemical Society Ref. [3]

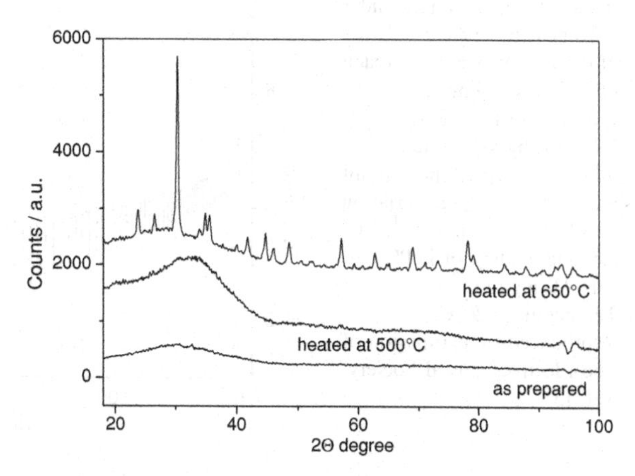

Fig. 5.2 X-ray powder diffraction patterns (Cu-Kα radiation) of the as-prepared sample and after firing at 500 and 650°C. Reproduced by permission of The Electrochemical Society Ref. [3]

for quartz-like materials of type MXO_4 (M = B, Al, Ga, Fe, Mn, and X = P, As) [5, 6].

Figure 5.2 shows the XRD patterns of the material as-prepared and after heating in air at various temperatures..

Upon heating at 500°C the residue product remains still amorphous. On further heating at 650°C the X-ray diffraction pattern shows a series of diffraction peaks, indicating that a transformation from amorphous to crystalline occurred at this temperature. The main peaks were attributed to crystalline anhydrous $FePO_4$, which has a hexagonal structure (JCPDS card No 29-0715). This phase resulted not isostructural with $FePO_4$ (heterosite) formed on lithium extraction from $LiFePO_4$ (triphylite) with ordered olivine-type structure [7].

Figure 5.3 is a SEM micrograph of the material obtained after precipitation. The material is characterized by a sponge-like structure where it is difficult to recognize any structural organization. The calculated material density was 2.24 g cm^{-3} that is lower than the crystalline material density, while the BET analysis gives rise to a specific surface area of 24.49 m^2 g^{-1}.

Fig. 5.3 SEM micrograph of the material as-obtained by precipitation. Reproduced by permission of The Electrochemical Society Ref. [3]

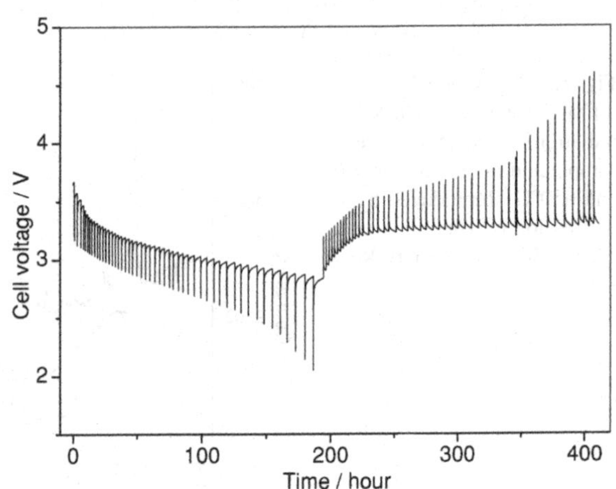

Fig. 5.4 Voltage profiles for the as-prepared sample under galvanostatic current pulses of 0.5 mA during 200 s (each one corresponding to $x = 0.02$ in Li_xFePO_4) followed by a potential relaxation step at open circuit until the cell voltage variation was less than 4 mV h^{-1}. The cathode loading of $FePO_4$ was 9.0 mg cm^{-2}. Temperature: 20°C. Reproduced by permission of The Electrochemical Society Ref. [3]

GITT was used to investigate the lithium intercalation–deintercalation process in the as-obtained $FePO_4$. For this experiment we used current pulses of 0.5 mA during 200 s (each one corresponding to $x = 0.02$ in Li_xFePO_4) followed by a potential relaxation step at open circuit until the cell voltage variation was less than 4 mV h^{-1}. Figure 5.4 illustrates the cell voltage as a function of time during the experiment. 1.0 lithium equivalent was inserted and reversibly de-intercalated into the material.

The cell over voltage was very high and was seen to increase at the end of the discharge process. This effect was even more evident at the end of the charge indicating a severe limitation to insert/remove lithium ions into/from the structure. These limitations could be related to slow diffusion of charge carriers in the active material.

The diffusion coefficient D was calculated according to Eq. 5.2 derived by Weppner and Huggins [8]:

$$D = 4/\pi(V_M/SF)^2\left[I^\circ(\delta E/\delta x)/\left(\delta E/\delta t^{1/2}\right)\right]^2 \quad \text{at } t << \tau \tag{5.2}$$

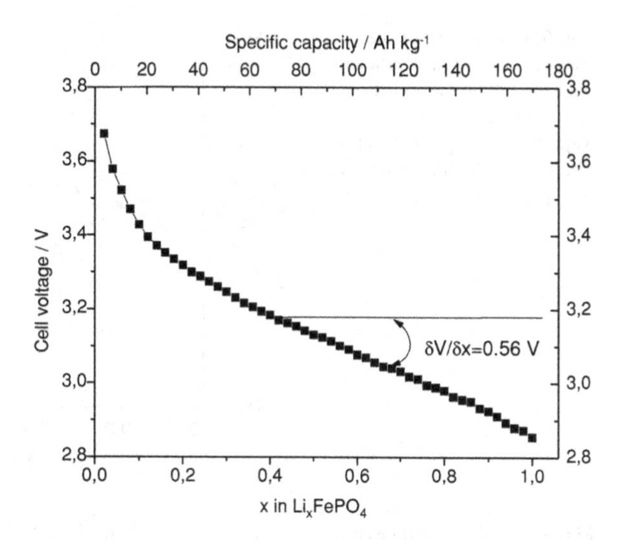

Fig. 5.5 The steady-state voltage versus Li in Li_xFePO_4 as a function of the stoichiometry x. The slope of the coulometric curve was found to be 0.56 V. The experimental conditions were the same as those of Fig. 5.4. Reproduced by permission of The Electrochemical Society Ref. [3]

where V_M is the phosphate molar volume (67.32 cm^3 mol^{-1}), S is the contact area between electrolyte and sample (1910 cm^2), F is the Faraday constant (96486 Q mol^{-1}), I° is the applied constant electric current (5×10^{-4} A), $\delta E/\delta x$ is the slope of the coulometric titration curve while $\delta E/\delta t^{1/2}$ is the slope of the transient voltage change.

The evolution of the quasi-equilibrium voltage versus x in Li_xFePO_4 is reported in Fig. 5.5. The data points were taken at the end of each relaxation step shown in Fig. 5.4. From an initial value of about 3.7 V, the quasi-equilibrium voltage sharply decreased with the first 0.1 equivalents of Li to about 3.4 V. After that, the slope of the open circuit voltage (OCV) curve versus x was found constant in the Li composition range up to 1.0, and equal to 0.56 V. The E–Eo versus $t^{1/2}$ plot was also found linear during the first 20 s. Figure 5.6 shows the diffusion coefficient as a function of x in Li_xFePO_4 obtained by substitution of the $\delta E/\delta t^{1/2}$ slopes, calculated for different values of x, in Eq. 5.2. The diffusion coefficient was found to range between 8.1×10^{-18} and 2.7×10^{-17}. These values are relatively low when compared to the diffusion coefficient of other 3 Volt active material used in lithium battery [9, 10].

The very low value of the diffusion coefficient can be related to the poor electronic conductivity of the material. The electronic conductivity in iron phosphates can be related to the electron hopping from the transition metal ions of lower valence states to those of higher valence states. Conductivity as low as 10^{-9} S cm^{-1} has been measured in iron phosphate glass systems at room temperature [11]. If the transference number of lithium ions is much larger than for the electrons, the lithium ion chemical diffusion coefficient is dependent on the component diffusion coefficient of electrons and its response to a compositional gradient will be very slow, in agreement with the measured values.

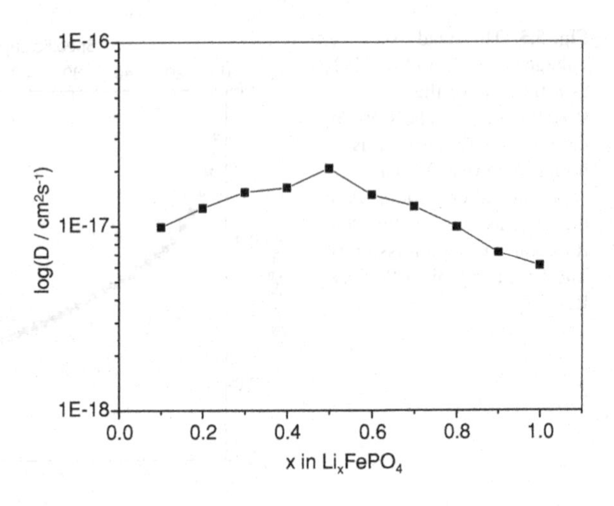

Fig. 5.6 Variation of the chemical diffusion coefficient over the whole composition range for Li_xFePO_4. Reproduced by permission of The Electrochemical Society Ref. [3]

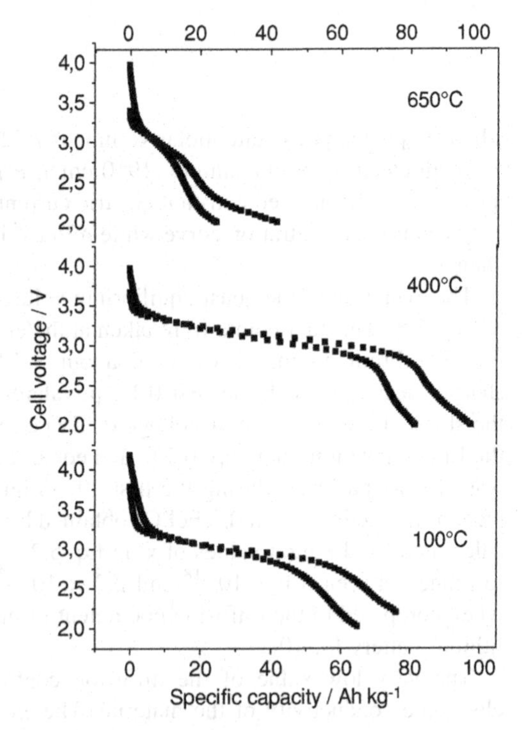

Fig. 5.7 Voltage profiles for samples fired at various temperatures. Discharge specific current was 170 A kg^{-1}. Temperature: 20°C. Reproduced by permission of The Electrochemical Society Ref. [3]

A first electrochemical test was performed to evaluate the effect of thermal treatments on the specific capacity exhibited by $FePO_4$. The material was tested after firing at 100, 400, and 650°C for 24 h.

Figure 5.7 shows the voltage profiles for the first and the second discharge cycle recorded under galvanostatic charge/discharge conditions at 170 A kg^{-1} rate. The capacity exhibited during the second cycle was less than the capacity

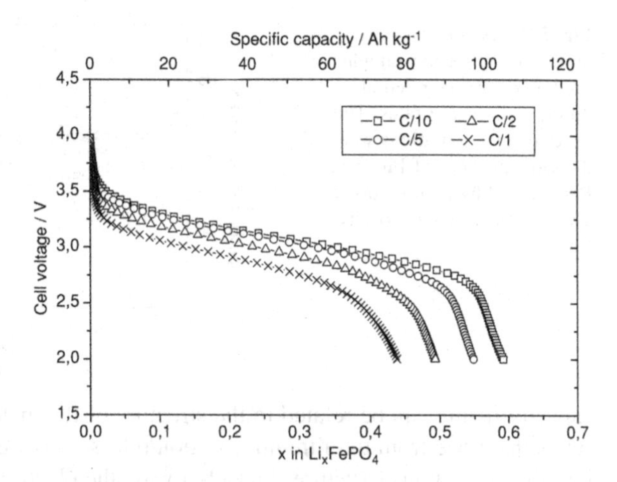

Fig. 5.8 Voltage profiles for the sample fired at 400°C recorded at different discharge rates. Charge current was 125 A kg^{-1}. A constant voltage of 4.0 V was applied at the end of the galvanostatic step until the current density was lower than 1/10th of the charge current. The cathode loading of $FePO_4$ was 7.6 mg cm^{-2}. Temperature: 20°C. Reproduced by permission of The Electrochemical Society Ref. [3]

delivered during the first intercalation step, due to the fact that part of the lithium inserted during the first discharge cycle was difficult to be de-intercalated. This result is in agreement with the larger over potential observed at the end of the charge in the GITT experiment. The specific capacity recorded during the second discharge cycle for the sample heated at 100°C was about 70 Ah kg^{-1}. The sample heated at 400°C showed a slightly increased capacity up to 80 Ah kg^{-1}. The increase in the specific capacity could be related to the increase of the active material content in the electrode due to water loss during the heating step. Finally, the sample heated at temperature higher than the crystallization temperature showed very poor electrochemical performance. The sample heated at 400°C was chosen to evaluate the capacity as a function of the discharge rate and the capacity retention as a function of prolonged cyclation. The cell was cycled galvanostatically under various discharge rate, namely 17, 34, 85, and 170 Ah kg^{-1}. The cut-off voltage was 2.0 V. The cell was always recharged with the same charge procedure, to assure identical initial conditions; a constant current step at 170 A kg^{-1} (C rate) until the voltage reached 4.0 V followed by a constant voltage step until the current lowered below 17 A kg^{-1} (C/10 rate). Figure 5.8 shows the voltage profiles recorded at various discharge rates.

At the lowest current density used (17 A kg^{-1}) the material was able to deliver a specific capacity of 108 Ah kg^{-1}. By increasing the current density the active material utilization decreased and about 80 Ah kg^{-1} was delivered when discharging the cell at 170 A kg^{-1}. The reversibility of the material at different discharge rates is showed in Fig. 5.9. The capacity retention was very good especially for the lower discharge rates. The moderate electrochemical performance of amorphous $FePO_4$ could be related to the low value of the diffusion coefficient that imposes a severe limitation to the active material utilization. The reduced utilization of the active material with increasing currents can also be related with kinetic limitation. The capacity fade, evaluated during cycling at the lowest discharge rate, was about 0.075% per cycle. Also in this case the low

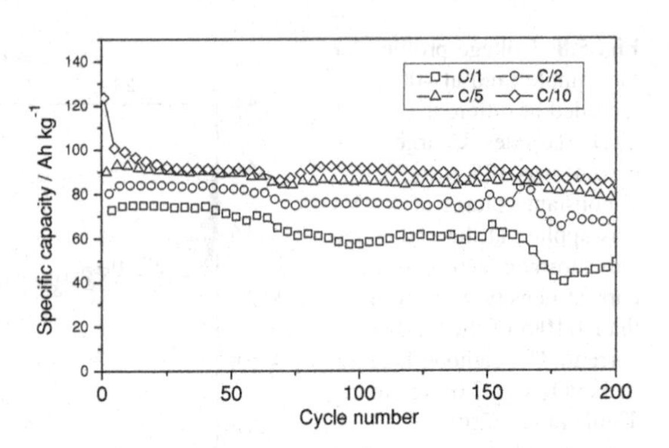

Fig. 5.9 Specific capacity upon cycling for the sample fired at 400°C recorded at different discharge rates. The experimental conditions were the same as those of Fig. 5.8. Reproduced by permission of The Electrochemical Society Ref. [3]

capacity fading can be related to the structure of the material. The fatigue and the stress that arise from the lithium insertion/release process are well accommodated into the amorphous structure. In such a way, the electrode structural changes that may cause contact loss and capacity fade upon cycling are reduced and the cycle life of the cathode enhanced.

5.3 Conclusions

Amorphous iron(III)phosphate was obtained by spontaneous precipitation from $Fe(NH_4)_2(SO_4)_2{\cdot}6H_2O$ and $NH_4H_2PO_4$, using hydrogen peroxide as an oxidizing agent. Chemical analysis and TG data suggested for the compound the following formula $FePO_4{\cdot}1.5H_2O$. The material was characterized as a cathode in non-aqueous lithium cells. GITT was used to evaluate the diffusion coefficient that was found to range from 8.1×10^{-18} to 2.7×10^{-17} cm^2 s^{-1}. These very low values were related to the poor electronic conductivity of the material. The electrochemical tests showed that the material was able to reversibly intercalate lithium. A specific capacity of about 100 Ah kg^{-1} was achieved by discharge the cell at 17 A kg^{-1}. The specific capacity was reduced at about 80% when discharging the cell at 170 A kg^{-1}. The moderate electrochemical performance can be related to the low value of the diffusion coefficient. The material showed good reversibility. More than 200 charge/discharge cycles were conducted with very satisfactory capacity retention. The capacity fade evaluated at 17 A kg^{-1} discharge rate was as low as 0.07% per cycle.

References

1. P.P. Prosini, L. Cianchi, G. Spina et al., Synthesis and characterization of amorphous $3Fe_2O_3{\bullet}2P_2O_5{\bullet}10H_2O$ and Its electrode performance in lithium batteries. J. Electrochem. Soc. **148**, A1125–A1129 (2001)
2. P.P. Prosini, M. Carewska, S. Loreti et al., Lithium iron oxide as alternative anode for li-ion batteries. Int. J. Inorg. Mater. **2**, 365–370 (2000)

3. P.P. Prosini, M. Lisi, S. Scaccia et al., Synthesis and characterization of amorphous hydrated $FePO_4$ and its electrode performance in lithium batteries. J. Electrochem. Soc. **149**, A297–A301 (2002)
4. N. Rajic, R. Gabrovsek, V. Kaucic, Thermal investigation of two FePO materials prepared in the presence of 1,2-diaminoethane. Thermochim. Acta. **359**, 119–122 (2000)
5. E. Philippot, A. Goiffon, A. Ibanez et al., Structure deformations and existence of the α-β transition in MXO_4 quartz-like materials. J. Solid State Chem. **110**, 356–362 (1994)
6. N. Aliouane, T. Badechet, Y. Gagou et al., Synthesis and phase transitions of iron phosphate. Ferroelectrics **241**, 255–262 (2000)
7. A.K. Padhi, K.S. Nanjundaswamy, C. Masquelier et al., Effect of structure on the Fe^{3+}/Fe^{2+} redox couple in iron phosphates. J. Electrochem. Soc. **144**, 1609–1613 (1997)
8. W. Weppner, R.A. Huggins, Determination of the kinetic parameters of mixed-conducting electrodes and application to the system Li_3Sb. J. Electrochem. Soc. **124**, 1569–1578 (1977)
9. L. Li, G. Pistoia, Secondary Li cells. II. Characteristics of lithiated manganese oxides synthesized from $LiNO_3$ and MnO_2. Solid State Ionics **47**, 241–249 (1991)
10. F. Coustier, S. Passerini, W.H. Smyrl, Dip-coated silver-doped V_2O_5 xerogels as host materials for lithium intercalation. Solid State Ionics **100**, 247–258 (1997)
11. K.K. Bahri, R.P. Tandon, M.C. Bansal, Effects of the additives on the electrical properties of iron oxide semiconducting glass. Eur. Phys. J. Ap. **4**, 291–296 (1998)

Chapter 6
Nano-Crystalline $LiFePO_4$

6.1 Preparation of Nano-Crystalline $LiFePO_4$

Amorphous $LiFePO_4$ was obtained by chemical lithiation of amorphous $FePO_4$ by using LiI as reducing agent. Amorphous $FePO_4$ was suspended in a 1 M solution of LiI in acetonitrile. The suspension was kept under agitation for 24 h, filtered on membrane filter (0.8 μm), washed several times with acetonitrile and dried under vacuum. Crystalline $LiFePO_4$ was obtained by heating the amorphous compound in a tubular furnace at 550°C for 1 h under reducing atmosphere (Ar/H_2).

The morphology of the samples was observed by scanning electron microscopy (SEM).

The material density was measured by using a helium picnometer (Accupyc 1330-Micromeritics). The specific surface area was measured by using a B.E.T. apparatus (Nova 2000-Quanta Chrome).

Composite cathode tapes were made by roll milling a mixture of 75 wt% active material, 5 wt% binder (Teflon, DuPont) and 20 wt% carbon (Super P, MMM carbon). Electrodes were punched in the form of discs typically with a diameter of 10 mm. The electrode weight ranged from 8.0 to 11.3 mg corresponding to an active material mass loading of 7.6–10.9 mg cm^{-2}. The electrodes were assembled in sealed cells formed by a polypropylene *T*-type pipe connector with three cylindrical stainless steel (SS316) current collectors. A lithium foil was used both as an anode and a reference electrode and a glass fibre was used as a separator. The cells were filled with ethylene carbonate/diethyl carbonate 1:1 $LiPF_6$ 1 M electrolyte solution. The cycling tests were carried out automatically by means of a battery cycler (Maccor 4000). Composite cathode preparation, cell assembly, test, and storage were performed in the dry room (R.H. $< 0.1\%$ at 20°C).

P. P. Prosini, *Iron Phosphate Materials as Cathodes for Lithium Batteries*,
DOI: 10.1007/978-0-85729-745-7_6,

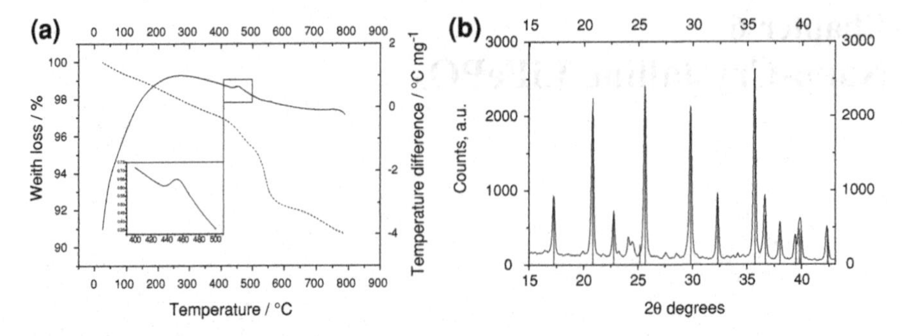

Fig. 6.1 **a** TG (.....), and DTA (—) curves for amorphous $LiFePO_4$ recorded over the temperature range from ambient to 800°C at a heating rate of 5°C min^{-1} in nitrogen atmosphere at 100 ml min^{-1} flow rate. **b** X-ray powder diffraction patterns (Cu–Kα radiation) of crystalline $LiFePO_4$ obtained by heating the amorphous precursor at 550°C for 1 h under reducing atmosphere (Ar/H_2). Reproduced by permission of The Electrochemical Society Ref. [2]

6.2 Physical and Electrochemical Characterization of Nano-Crystalline Lithium Iron Phosphate

In the previous chapter it was shown that amorphous $FePO_4$ can be obtained by following a solution-based approach [1]. In this chapter it is described how to use this material to obtain amorphous nano-sized $LiFePO_4$ and nano-crystalline $LiFePO_4$ with ordered olivine-type structure [2]. Amorphous $LiFePO_4$ was obtained by chemical lithiation of amorphous $FePO_4$ by using LiI as reducing agent according with Eq. 6.1:

$$FePO_4 + LiI \rightarrow LiFePO_4 + 0.5\,I_2 \tag{6.1}$$

The chemical analysis of the lithiated compound gives rise to a molar ratio Li_t:Fe_t:P_t = 1:1:1, suggesting the amorphous $FePO_4$ was completely lithiated.

The TG/DTA curves of the lithiated compound are displayed in Fig. 6.1a. Over the temperature range from ambient to 550°C there is no appreciable weight loss in the TG curve. The corresponding DTA curve shows an exothermic peak at 470°C that is related to the crystallization of the compound.

Figure 6.1b shows the X-ray diffraction patterns of the material after heating in argon/H_2 at 550°C for 1.0 h. The X-ray diffraction pattern shows a series of diffraction peaks, indicating that a transformation from amorphous to crystalline phase occurred at this temperature. The main peaks were attributed to crystalline $LiFePO_4$, which has an olivine structure (JCPDS card No 42-0580). Figure 6.2 shows SEM micrographs of the lithiated materials before and after the heating treatment. The materials are characterized by a globular structure with a grain size of about 100–150 nm. The B.E.T. analysis gives rise to a specific surface area of 8.95 $m^2\ g^{-1}$ for the crystalline material.

Figure 6.3a shows the voltage profiles as a function of the specific capacity for several discharge rates. The cell was discharged galvanostatically under different

Fig. 6.2 **a** SEM micrograph of amorphous $LiFePO_4$ as obtained after chemical lithiation and **b** after heating the amorphous precursor at 550°C for 1 h under reducing atmosphere (Ar/H_2). Reproduced by permission of The Electrochemical Society Ref. [2]

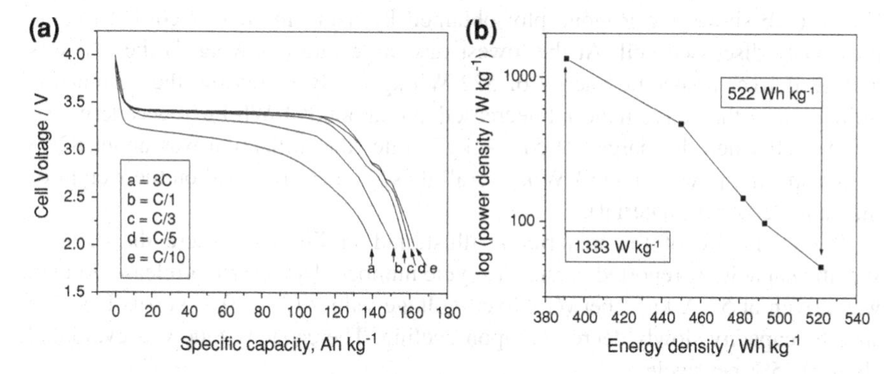

Fig. 6.3 **a** Voltage profiles for a $LiFePO_4$ fired at 550°C for 1 h recorded at different discharge rate. Specific charge current was 17 A kg^{-1}. The cathode loading of $LiFePO_4$ was 7.6 mg cm^{-2}. Temperature: 20°C. **b** Ragone plot. Reproduced by permission of The Electrochemical Society Ref. [2]

specific currents, ranged from 17 up to 510 A kg^{-1}. The cut-off voltage was 2.0 V. The cell was always recharged at the same specific current (17 A kg^{-1}), to assure identical initial conditions. At the lowest discharge current used (17 A kg^{-1}), the cell was able to deliver a specific capacity of 162 Ah kg^{-1}, based on the active material weight, corresponding to a discharge time of about 10 h. By increasing the current density the utilization of the active material decreased, and about 140 Ah kg^{-1} was delivered in about 0.29 h at a specific current 30 times higher. This result appears very good when compared with iron phosphate synthesized by traditional solid-state chemistry, in which an increase in the discharge current results in a severe capacity fading [3].

The excellent performance of the compound can be ascribed to the small particle size achieved by the solution-based synthesis, followed by low temperature crystallization, compared to the procedures reported in the literature.

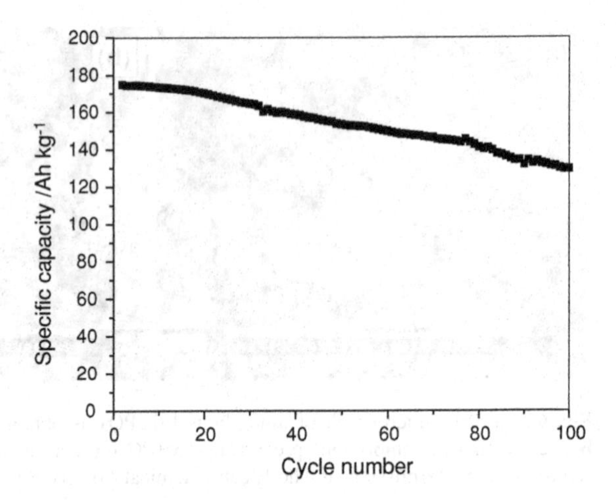

Fig. 6.4 Specific capacity upon discharge; the insertion/release process was driven at a specific current of 57 A kg^{-1}. The cathode loading of $LiFePO_4$ was 10.88 mg cm^{-2}. Reproduced by permission of The Electrochemical Society Ref. [2]

Figure 6.3b shows the Ragone plot obtained by using the data recorded for the previously discussed cell. At the lowest discharge rate (17 A kg^{-1}) the cell was able to deliver a specific energy of 522 Wh kg^{-1}. By increasing the current the utilization of the active material decreased and about 391 Wh kg^{-1} was delivered by the cell when discharged at 510 A kg^{-1} rate (the utilization was about 82%) with a specific power of 1333 W kg^{-1} (all these values are based on the weight of the cathode active material).

The cycle life of the material is illustrated in Fig. 6.4, where the cathode specific capacity is reported versus the cycle number. The insertion/release process was driven at 57 A kg^{-1} between fixed voltage values (2.0/4.0 V versus Li). The specific capacity slowly decreased upon cycling. The capacity fade was evaluated about 0.25% per cycle.

6.3 Conclusions

Nano-crystalline $LiFePO_4$ was synthetized by heating amorphous $LiFePO_4$ obtained by chemical lithiation of $FePO_4$. The small particle size achieved by the solution-based synthesis was shown to enhance the electrochemical performance of the material. Nano-crystalline $LiFePO_4$ discharged at 17 A kg^{-1} rate exhibited a specific energy of 522 Wh kg^{-1}. A capacity fading of about 0.25% per cycle affected the cell upon cycling.

References

1. P.P. Prosini, M. Lisi, S. Scaccia et al., Synthesis and characterization of amorphous hydrated $FePO_4$ and its electrode performance in lithium batteries. J. Electrochem. Soc. **149**, A297–A301 (2002)

2. P.P. Prosini, M. Carewska, S. Scaccia et al., A new synthetic route for preparing $LiFePO_4$ with enhanced electrochemical performance. J. Electrochem. Soc. **149**, A886–A890 (2002)
3. A.K. Padhi, K.S. Nanjundaswamy, J.B. Goodenough, Phospho-olivines as positive-electrode materials for rechargeable lithium batteries. J. Electrochem. Soc. **144**, 1188–1194 (1997)

Chapter 7
Long-Term Cyclability of Nano-Crystalline $LiFePO_4$

7.1 Effect of Firing Time on Electrochemical Performance of Nano-Crystalline $LiFePO_4$

In the previous chapter it was shown that nano-crystalline $LiFePO_4$ can be prepared by heating at 550°C for 1 h amorphous nano-sized $LiFePO_4$ [1]. In this chapter, to evaluate the effect of firing time on the electrochemical performance, the amorphous precursor was heated at 550°C for different periods of time and the obtained materials tested as a cathode in lithium batteries [2]. Figure 7.1 shows the X-ray diffraction patterns of the amorphous material after heating in argon/H_2 at 550°C for 1 and 5 h. For comparison, the same figure shows the position and the relative intensity of the peaks of a crystalline $LiFePO_4$ sample (JCPDS card No 42-0580). It should be noted that 1 h is sufficient to crystallize the material. The grain-size (D) was calculated using the Scherrer formula: $\beta \cos(\theta) = k\lambda/D$, where β is the full-width-at-half-maximum length of the diffraction peak on a 2θ scale and k is a constant here close to unity. No great variation in the crystalline grain-size was observed by changing the heating treatment. The mean value of D computed from the (120), (111), (200), and (131) diffraction peaks (the best resolved in the diffractograms) ranges from 85 nm (1 h heated sample) to 90 nm (5 h heated sample). Figure 7.2 shows the specific surface area of samples heated at 550°C as a function of firing time. The specific surface area decreased by raising the firing time: this can be related to the partial coalescence of the $LiFePO_4$ particles during the heat treatment.

Figure 7.3a and b are SEM micrographs of the materials after heat treatments for 1 and 5 h. Both are characterized by a globular structure with grain-sizes c.a. 100–150 nm. SEM micrographs confirm that the material annealed for a longer time showed a partial coalescence of the grains.

Figure 7.4a shows the voltage profile as a function of the specific capacity for the material annealed for 5 h.

The cell was discharged galvanostatically under different specific currents ranging from 17 to 510 A kg^{-1}. The cut-off voltage was 2.0 V. The cell was

P. P. Prosini, *Iron Phosphate Materials as Cathodes for Lithium Batteries*,
DOI: 10.1007/978-0-85729-745-7_7,

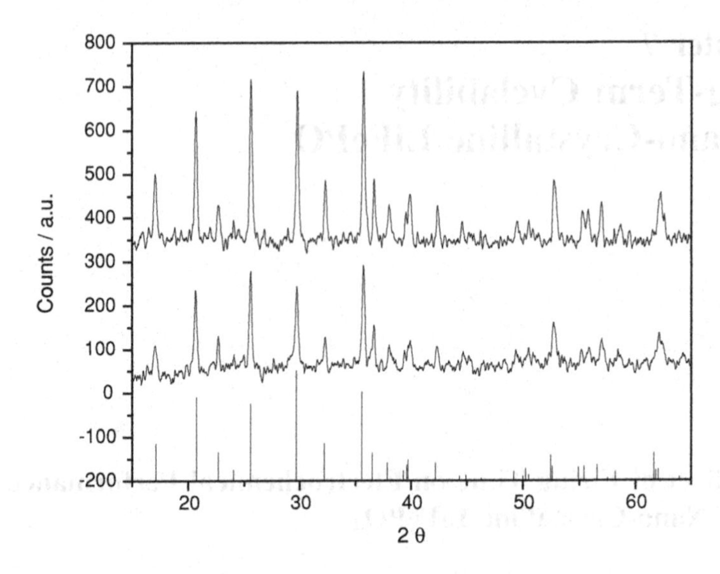

Fig. 7.1 X-ray powder diffraction patterns (Cu–Kα radiation) of crystalline $LiFePO_4$ obtained by heating the amorphous precursor at 550°C for 1 and 5 h (lower and upper curves, respectively) under reducing atmosphere (Ar/H_2). The position and the relative intensity of the peaks of crystalline $LiFePO_4$ are reported for comparison. Reproduced by permission of Elsevier Ref. [2]

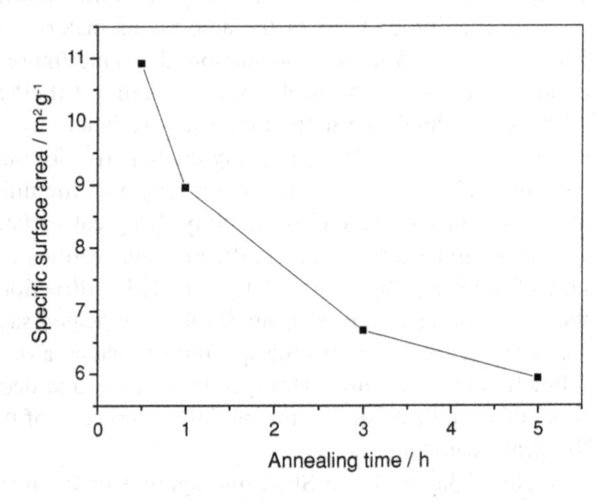

Fig. 7.2 Specific surface area of samples fired for different periods of time at 550°C as a function of the firing time. Reproduced by permission of Elsevier Ref. [2]

always recharged at the same specific current (17 A kg^{-1}) to assure identical initial conditions. At the lowest discharge current used (17 A kg^{-1}), the cell was able to deliver a specific capacity of 155 Ah kg^{-1}, based on the weight of active material, corresponding to a discharge time of about 10 h. By increasing the current density, the utilization of the active material decreased: c.a. 133 Ah kg^{-1} was delivered in 0.29 h at a specific current 30 times larger. Figure 7.4b shows the Ragone plot for three different thermally-treated samples. During the first set of

Fig. 7.3 **a** SEM micrographs of crystalline $LiFePO_4$ obtained by heating the amorphous precursor at 550°C for 1 h and **b** 5 h under reducing atmosphere (Ar/H_2). Reproduced by permission of Elsevier Ref. [2]

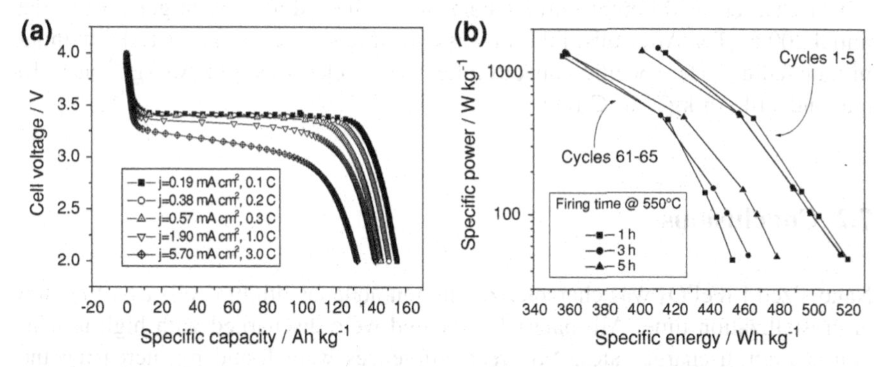

Fig. 7.4 **a** Voltage profiles for a $LiFePO_4$ fired at 500°C for 5 h recorded at different discharge rates. Specific charge current was 17 A kg^{-1}. The cathode loading of $LiFePO_4$ was 12.38 mg cm^{-2}. Temperature: 20°C. **b** Ragone plot for different thermally-treated $LiFePO_4$ samples. The experimental conditions are the same as in Fig. 7.4. The cathode loading of $LiFePO_4$ ranged from 12.20 to 12.40 mg cm^{-2}. Reproduced by permission of Elsevier Ref. [2]

measurements, all samples showed identical behavior. A specific energy exceeding 515 Wh kg^{-1} was calculated for all samples in the lowest current-density cycle. At the highest current density, the samples showed a power density exceeding 1380 W kg^{-1}, while maintaining over than 80% of the specific energy (420 Wh kg^{-1}). After 60 cycles, the sample heated for 5 h showed the best electrochemical performance, delivering a specific energy of about 480 Wh kg^{-1}. This effect disappeared progressively when increasing the discharge current and, at the highest discharge current, all materials showed the same electrochemical performance. The specific energy was about 360 Wh kg^{-1}, with a power density exceeding 1300 W kg^{-1}. All samples showed a pronounced capacity decline during the first cycles. Figure 7.5 shows the cycle life of the material annealed for

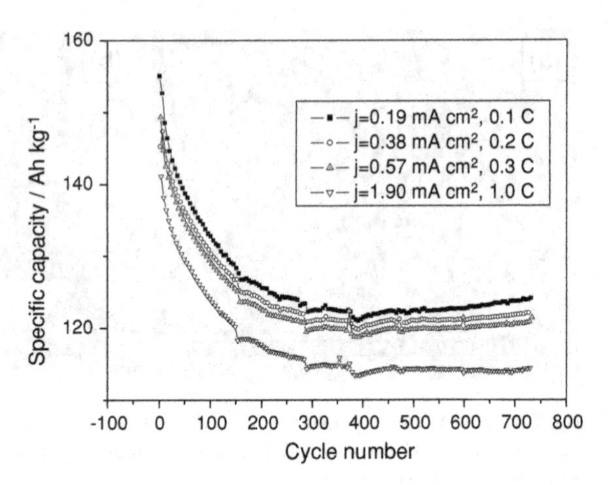

Fig. 7.5 Specific capacity upon discharge; the insertion/release process was driven at different specific currents. Specific charge current was 17 A kg^{-1}. The cathode loading of $LiFePO_4$ was 12.20 mg cm^{-2}. Temperature: 20°C. Reproduced by permission of Elsevier Ref. [2]

3 h discharged at different current densities. The fading was larger during the initial 200 cycles. After this, lithium was cycled reversibly in the material without further fading. The specific capacity after 700 cycles was 124 Ah kg^{-1} at C/10 rate and 114 Ah kg^{-1} at C rate.

7.2 Conclusions

Nano-sized $LiFePO_4$ was characterized in non-aqueous lithium cells as a function of crystallization time. All materials prepared were discharged with high utilization at high-discharge rates. No great differences were found by increasing the firing time; 1 h was enough to prepare a material with good electrochemical performance. A capacity fading affected all the materials but, after 200 deep-discharge cycles, the capacity fading was drastically reduced and the cells were able to cycle 500 times without further fading. The initial fading could be related to contact losses between the conductive binder and the active material particles or $LiFePO_4$ structural variations. The contact losses could result from volume variations occurring in $LiFePO_4$ during lithium insertion/extraction.

References

1. P.P. Prosini, M. Carewska, S. Scaccia et al., A new synthetic route for preparing $LiFePO_4$ with enhanced electrochemical performance. J. Electrochem. Soc. **149**, A886–A890 (2002)
2. P.P. Prosini, M. Carewska, S. Scaccia et al., Long-term cyclability of nanostructured $LiFePO_4$. Electrochim. Acta **48**, 4205–4211 (2003)

Chapter 8
Factors Affecting the Rate Performance of $LiFePO_4$

8.1 Preparation of the Composite Cathodes

Composite $LiFePO_4/C$ was obtained by heating amorphous $LiFePO_4$ and carbon in a tubular furnace at 550°C under reducing atmosphere ($Ar/H_2 = 95/5$) for 1 h. After the heat treatment, the material was allowed to cool to room temperature [1]. The material was characterized by X-ray powder diffraction analysis (Philips PW 3710 diffractometer) using Cu–Kα radiation.

To increase the electric conductivity the active material and the carbon (KJB, Akzo Nobel) were mixed in a mortar for 5 min. An extensive mixing procedure consisting of blending the powders for 15 min was also performed. The binder (Teflon, DuPont) was added and the blend was mixed to obtain a plastic-like material. The quantity of binder was kept constant (10 wt%) while the active material/carbon ratio was varied to prepare mixtures with different carbon content, namely 10, 15, and 20 wt%. Composite cathode tapes were made by roll milling the so-obtained mixtures.

Composite films morphology and composition were studied by a scanning electron microscope (SEM) Jeol JSM-5510LV with an energy dispersive X-ray analysis (EDS) IXRF EDS-2000 System. The specimens were directly mounted onto conductive carbon double face tape, which was previously mounted on a slab. The conditions were: accelerating voltage 25 kV, spot size 31 and working distance 21 mm.

Electrodes were punched in the form of discs, typically with a diameter of 8 mm. The electrode weight ranged from 7.0 to 10.0 mg, corresponding to an active material mass loading of 10.9–15.3 mg cm^{-2}. Polypropylene T-type pipe connectors with three cylindrical stainless steel (SS316) current collectors were used as cells. Lithium was used both as an anode and a reference electrode and a glass fibre was used as a separator. The cells were filled with a 1 M solution of $LiPF_6$ in ethylene carbonate/diethyl carbonate (1/1). The cycling tests were automated with a battery cycler (Maccor 4000). The impedance was measured with a frequency response analyzer (FRA, Solartron) interfaced with a personal

P. P. Prosini, *Iron Phosphate Materials as Cathodes for Lithium Batteries*,
DOI: 10.1007/978-0-85729-745-7_8, © Springer-Verlag London Limited 2011

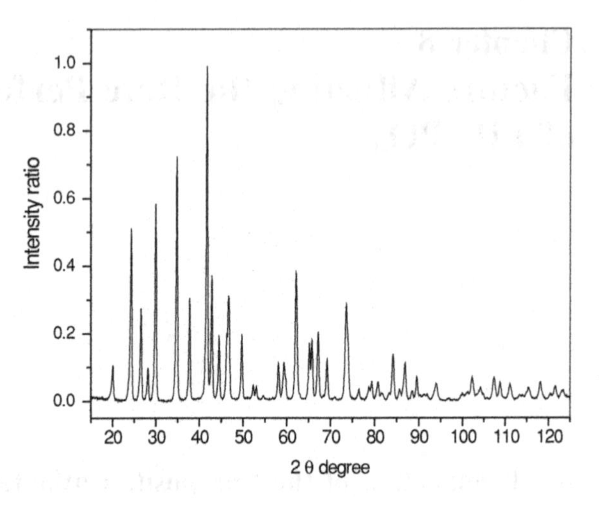

Fig. 8.1 X-ray powder diffraction patterns (Cu–Kα radiation) of crystalline $LiFePO_4$ obtained by heating the amorphous precursor at 550°C for 1 h under reducing atmosphere (Ar/H_2). Reproduced by permission of Elsevier Ref. [2]

computer, over a frequency range from 10 MHz to 0.01 Hz. with a 10 mV amplitude. Composite cathode preparation, cell assembly, tests and storage were performed in a dry room (R.H. < 0.1% at 20°C).

8.2 Effect of Carbon Content on the Electrochemical Performance of Nano-Crystalline $LiFePO_4$

Figure 8.1 shows the X-ray diffraction patterns of the active material. The comparison with published spectra of Li–Fe–P–O reveals the presence of single phase $LiFePO_4$ (40-1499 card JCPDS Data Base). The grain size (D) was calculated with the Scherrer formula: $\beta \cos(\theta) = k\lambda/D$, where β is the full-width-at-half-maximum length of the diffraction peak on a 2θ scale and k is a constant here close to unity. The mean value of D is about 100 nm. Figure 8.2a–c shows the SEM images and the corresponding EDS element (Fe, C, and F) maps of composite films containing 10, 15, and 20 wt% carbon, 10 wt% Teflon balanced with $LiFePO_4$, respectively. All composite films show a uniform fine-grained microstructure with particle size in the range 100–200 nm. By increasing the carbon content, the surface microstructure of the film becomes denser.

The distribution area for iron and carbon is homogeneous, whereas an evident non-uniform distribution of Teflon in the film with the lowest carbon content is observed (Fig. 8.2a). The composite films were used as cathodes in non-aqueous lithium cells. The cells were discharged galvanostatically under different specific currents ranging from C/10 to 10C. The cut-off voltage was 2.0 V. The cells were always recharged at the same specific current (C/10, 17 A kg^{-1}) to assure identical initial conditions. Figure 8.3a shows the log/log plot reporting the energy ratio (E_D/E_M) as a function of the discharge rate for three cathodes with different

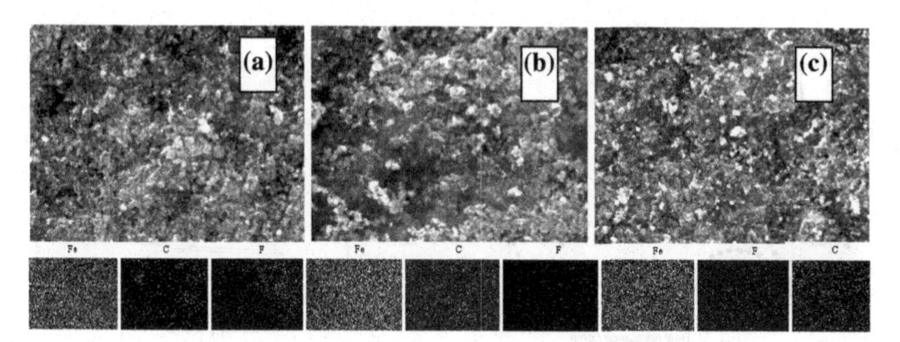

Fig. 8.2 SEM images and corresponding EDS (Fe, C, and F) maps of composite films containing **a** 10, **b** 15, and **c** 20 wt% carbon. Reproduced by permission of Elsevier Ref. [2]

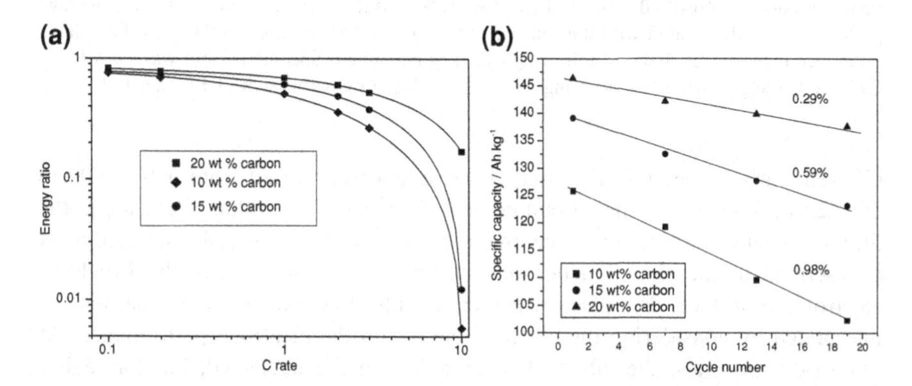

Fig. 8.3 **a** Log/log plot reporting the energy ratio (E_D/E_M) as a function of the discharge rate for three different cathodes prepared with the carbon content reported in the figure. The maximum energy (E_M) is the theoretical one. E_D is the energy delivered at the specified discharge rate. Cathode loading of $LiFePO_4$: 11.7 mg cm^{-2}. **b** Specific capacity upon discharge (C/10 discharge rate) for the cathodes previously described. The carbon content and the capacity fading (calculated as a percentage per cycle) are reported in the figure. Specific charge current: 17 A kg^{-1}. Reproduced by permission of Elsevier Ref. [2]

carbon content. E_D is the energy delivered at the specified discharge rate. The maximum delivered energy (E_M) is the theoretical one (580 Wh kg^{-1}).

As expected the E_D/E_M ratio decreases by increasing the discharge current and this effect is more pronounced for low carbon content cathodes. Fading (Fig. 8.3b) was observed for a 1C discharge rate and becomes progressively greater for larger discharge rates. At a 10C discharge rate, only the cell with 20 wt% carbon shows significant performance. It is interesting to note that the delivered specific capacity is lower than the theoretical one (170 Ah kg^{-1}) and that it decreases by decreasing the carbon content. The capacity fading, reported in the figure as percentage per cycle, progressively increases.

Figure 8.4a shows a comparison of the impedance spectra after one charge/discharge test recorded in a two electrode cell configuration for cathodes with

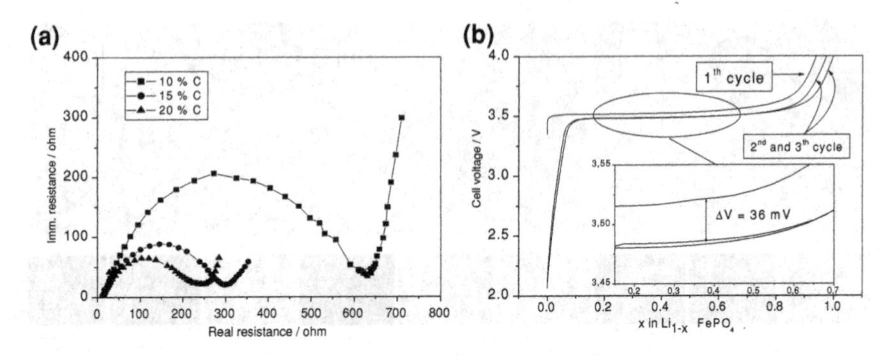

Fig. 8.4 **a** Comparison of the impedance spectra after one charge/discharge test recorded in a two electrodes cell configuration for cathodes with different carbon content as reported in the figure. Frequency range: 0.1 Hz–10 kHz. Electrode surface area: 0.5 cm^2. **b** Charge voltage profiles for a cell prepared with the cathode containing 20 wt% carbon. In the inset the voltage difference between the first and following cycles is magnified. Specific current was 17 A kg^{-1}. Cathode loading of $LiFePO_4$: 15.3 mg cm^{-2}. Reproduced by permission of Elsevier Ref. [2]

different carbon content. The total cell resistance decreased from 600 to 300 to 250 Ω as a function of the carbon content. Given the low value of the R_{ct} at the lithium metal/electrolyte interface, the large value of the total cell resistance must be ascribed to the slow kinetics of lithium ion incorporation at the $LiFePO_4$/electrolyte interface. The electrochemical results shown in Fig. 8.3a can now be interpreted by considering the effect of the R_{ct} on the electrode performance. At low discharge rates, the effect of a large R_{ct} can be neglected, but for higher discharge rates the R_{ct} is mainly responsible for the voltage drop causing a sudden decrease in the electrochemical performance. For the described cell, a 10C rate corresponds to a current of 7 mA, that when flowing in the cell with a total resistance of 250 Ω, results in a voltage drop of 1.78 V which is slightly lower than the working electrochemical window (2 V, from 4 to 2 V). At a 10C rate, the voltage drop due to R_{ct} for the other two cathodes is higher than the working electrochemical window, accounting for the very poor energy delivered. The capacity fading can also be related to the carbon content. Cells with higher carbon content showed a reduced fading. The fading may be caused by a loss of contact between the active material particles and the conductive filler.

Upon charging, the electronic conductor particles are forced to move following the active material volume variations and the contact may be lost between the electric conductor particles and the active material when the volume decreases upon discharging. Increasing the amount of carbon increases the probability of preserving the contact upon cycling. Figure 8.4b shows the voltage profiles upon charging for the cathode containing 20 wt% carbon. The first cycle charging voltage is higher than in the following cycles (the second and third cycle are reported for comparison). The inset in the figure shows an enlargement of the voltage profiles. The voltage difference between the first and following charge cycles was about 36 mV. All samples examined showed the same behavior with slight variations in the voltage gap. For the cell described in Fig. 8.3, the charging was driven at 0.131 mA (C/10 rate) and from the Ohm's Law a resistance of about 270 Ω can be calculated. This value is very close to the R_{ct} value as obtained from IS measurements (about 250 Ω). The same correlation was found for other cathodes, suggesting that the decrease of the cell voltage after the first cycle can be related to the decrease of the R_{ct} at the $LiFePO_4$/electrolyte interface. The reduction of the R_{ct} could be related to the formation of an ionically conducting surface film as indicated by photoelectron spectroscopy [3]. The formation of a new fresh lithium surface at the lithium metal/electrolyte interface starts just after the current is applied to the cell and is completed few minutes after the starting [4]. For this reason the decrease of the R_{ct} at the lithium metal/electrolyte interface does not contribute to the first cycle voltage decrease.

All cathodes showed extremely high values of the R_{ct}, making them unsuitable for high power applications. The total impedance for the best cell tested was about 250 Ω. Considering that the carbon content in all of the cathodes is larger than the conductivity threshold [4], we tried to decrease the R_{ct} by increasing the mixing time from 5 to 15 min.

Impedance spectroscopy using both a two (2EI) or a three electrodes (3EI) cell configuration was conducted on fresh cells (before) as well as in cycled cells (after the test cycles). The spectra recorded for the sample containing 20 wt% of carbon are reported in Fig. 8.5a–d. In the 2EI cell configuration the two semicircles related to the R_{ct} at the $LiFePO_4$/electrolyte and at the lithium metal/electrolyte interfaces are overlapped giving rise to a depressed semicircle from which a total cell resistance larger than 500 Ω is found. Thus, the lithium metal/electrolyte contribution to the total cell resistance is about 200 Ω. The high value of the R_{ct} at the lithium metal/electrolyte interface may be related to a passive layer intrinsically present on the lithium foil or formed when in contact with the electrolyte. Figure 8.5b shows the impedance spectra before and after one charge/discharge step. The R_{ct} related to the $LiFePO_4$/electrolyte interface is seen to decrease by about 100 Ω after the passage of current. The total cell resistance shows a large decrease as shown in Fig. 8.5c, decreasing to about 300 Ω. This may be related to the destruction of the passivation layer on the lithium metal surface. It is well known that the passivation layer can be easily destroyed by applying a constant current between the electrodes [5]. In such a way, a new fresh lithium surface is formed and the interface resistance is reduced to a very low value. Figure 8.5d shows a comparison of the impedance spectra after one charge/discharge step recorded in a 2 or 3EI cell configuration. The R_{ct} at the $LiFePO_4$/electrolyte interface in the 3EI cell configuration is found to be about 200 Ω. In the 2EI cell configuration, the two semicircles related to the R_{ct} at the $LiFePO_4$/electrolyte and lithium metal/electrolyte interfaces are well resolved and a R_{ct} at the lithium metal/electrolyte of about 40 Ω is found. Samples with lower carbon content showed the same "activation" behavior upon cycling with the difference that the resistance values are larger with respect to the cathode discussed above.

Figure 8.6 shows the SEM micrograph and iron, carbon, and fluorine EDS maps of the composite films containing 20 wt% carbon prepared with the longer

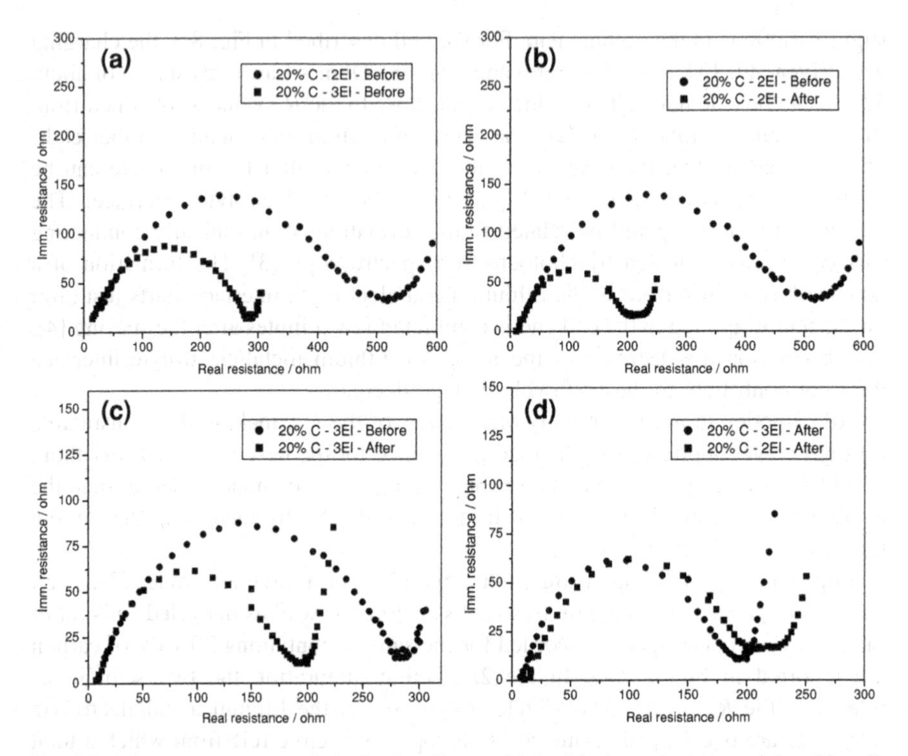

Fig. 8.5 **a** Impedance spectra for a fresh cell assembled with the cathode containing 20 wt% carbon in 2 and 3EI cell configuration. **b** Comparison of the impedance spectra before and after one charge/discharge step in a 2EI cell configuration. **c** Comparison of the impedance spectra before and after one charge/discharge step recorded in a 3EI cell configuration. **d** Impedance spectra after one charge/discharge step in 2 and 3EI cell configuration. Frequency range: 0.1 Hz–10 kHz. Electrode surface area: 0.5 cm^2. Reproduced by permission of Elsevier Ref. [2]

mixing procedure. The microstructure appears very dense with uniform distributions of carbon and Teflon.

Figure 8.7a shows the impedance spectra for the cathode prepared with the improved procedure. By increasing the mixing time, a large decrease of the R_{ct} was attained and the cell impedance at medium frequency was reduced to 38 Ω. The contribution of R_{el} to the total resistance, evaluated at high frequency, was about 8 Ω. The strong reduction of the total cell impedance should result in a better material utilization especially at high discharge regimes. Figure 8.7b shows the Ragone plot for the cathodes containing 20 wt% carbon, prepared with and without the extensive mixing procedure.

The cathode prepared with the longer mixing procedure showed the best electrochemical performance both at low and high discharge rate. At the lowest discharge rate (C/10), the optimized cathode delivered a specific energy close to the theoretical one. By increasing the discharge current, the utilization of the active material decreased. However, about 350 Wh kg^{-1} was delivered by the cell when

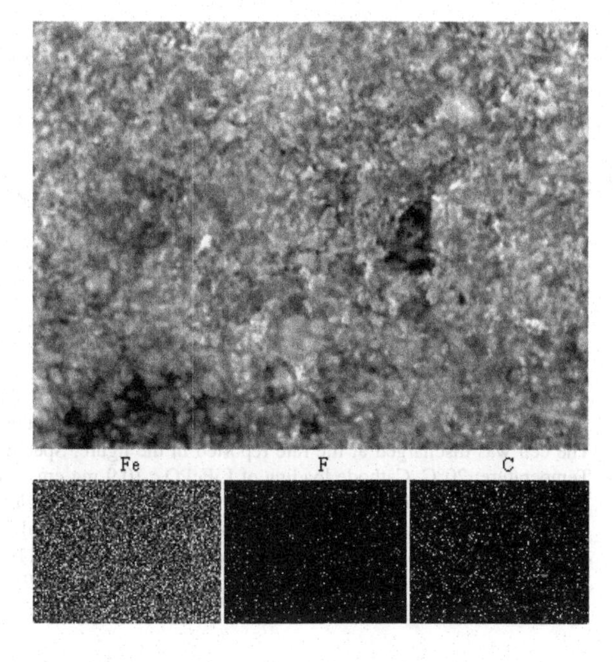

Fig. 8.6 SEM image and EDS (Fe, C, and F) maps of a composite film containing 20 wt% carbon prepared with the longer time mixing procedure. Reproduced by permission of Elsevier Ref. [2]

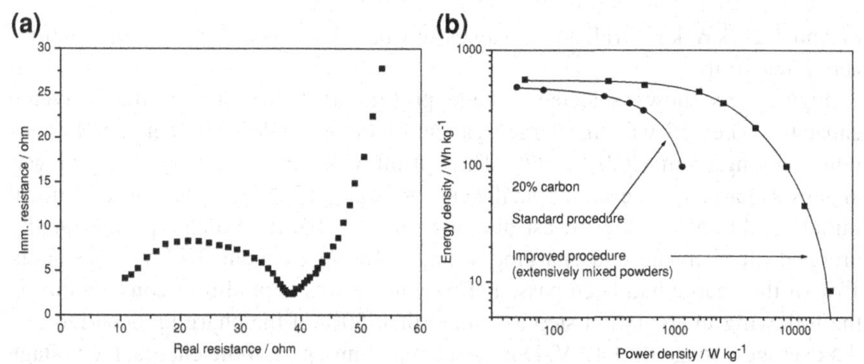

Fig. 8.7 **a** Impedance spectra for a cell assembled with the cathode prepared with the longer blending procedure. The spectrum was recorded in a 2EI cell configuration. Cathode carbon content was 20 wt%. Frequency range: 0.1 Hz–10 kHz. Electrode surface area: 0.5 cm^2. **b** Comparison of the Ragone plot for the cathodes containing 20 wt% carbon and prepared with different blending procedures. Cathode loading of $LiFePO_4$: ranged from 11.4 mg cm^{-2} (standard procedure) to 10.9 mg cm^{-2} (longer time procedure). Reproduced by permission of Elsevier Ref. [2]

discharged at 3C rate, corresponding to an active material utilization of 60% and a specific power of 2.5 kW kg^{-1}. About 215 Wh kg^{-1} was delivered when discharged at 5C rate, corresponding to an active material utilization of about 40% and a specific power larger than 4.5 kW kg^{-1}. The cell was able to sustain current flows as high as 50 mA cm^{-2} for several seconds with a maximum output power

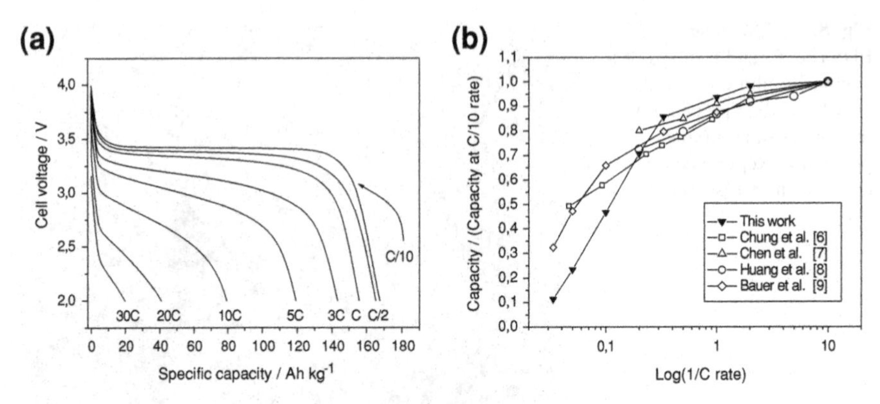

Fig. 8.8 **a** Discharge voltage profiles for a cell prepared with the cathode described in Fig. 8.7. The cell was discharged at the rate reported in the figure. Specific charge current: 170 A kg^{-1}. Temperature: 20°C. Cathode loading of $LiFePO_4$: 10.9 mg cm^{-2}. **b** Relative specific capacity of undoped $LiFePO_4$ (10.9 mg cm^{-2}) 20 wt% carbon (*fill triangle*), material reported by Chung et al. [6] (2.5 mg cm^{-2}) 10 wt% carbon (*square*), material reported by Chen et al. [7] (10 mg cm^{-2}) > 7% carbon (*empty triangle*), material reported by Huang et al. [8] (5 mg cm^{-2}) 20 wt% carbon (*circle*), and material reported by Bauer et al. [9] (7.1 mg cm^{-2}) 20 wt% carbon (*diamond*). Reproduced by permission of Elsevier Ref. [2]

of about 20 kW kg^{-1} (all these values are based on the weight of the cathode active material).

Figure 8.8a shows selected voltage profiles as a function of the delivered capacity. The cell was discharged galvanostatically under different specific currents ranging from C/10 to 30C. The cut-off voltage was 2.0 V. The cell was always recharged at the same specific current (C/10, 17 A kg^{-1}) to assure identical initial conditions. At the lowest discharge rate (C/10) the voltage profile rapidly dropped off from the end charging voltage (4.0 V) to about 3.43 V, after about 15% of the charge had been passed. This voltage was kept almost constant during the following intercalation step and more than 50% of the charging occurred at a flat average voltage of 3.42 V. During the remaining part of the charge, the voltage quickly dropped off to reach the end charging voltage (2.0 V). By increasing the discharge current, two effects can clearly be observed in the flat voltage region: a decrease of the average discharge voltage and an increase of the cell voltage slope. For currents lower than a 3C rate, a slight capacity fading was observed. At higher discharge regimes, a progressive capacity decline with increasing currents was observed. Figure 8.8b shows the cell performance as a function of the logarithm of the inverse discharge current compared to other doped or carbon coated $LiFePO_4$-based cathodes.

The electrochemical performance is reported in terms of capacity ratio between the delivered capacity at a specified rate and the capacity delivered at C/10 rate. For currents lower than 3C, all materials showed the same electrochemical properties in terms of capacity retention. Current higher than 3C were investigated by Chung [6] and Bauer [9]. The carbon coated material reported by Bauer shows

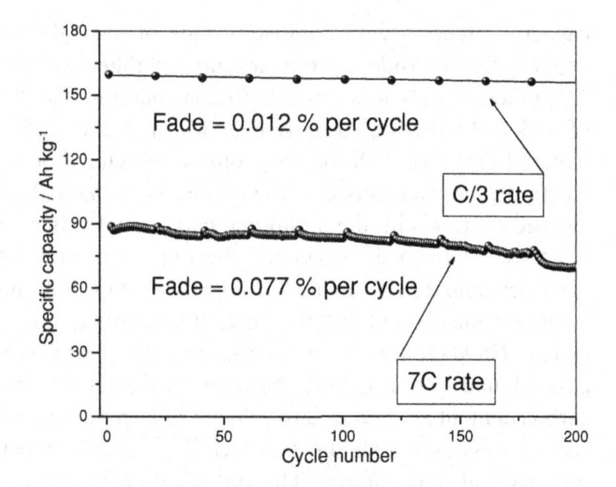

Fig. 8.9 Specific capacity upon discharge in which the insertion/release process was driven at the two specific current rates reported in the figure. Specific charge current: 170 A kg^{-1} with a top-off at 4.0 V. This voltage was applied to the cells until the current was decreased to 1/10th of its initial value. Cathode loading of $LiFePO_4$: 12.7 mg cm^{-2}. Reproduced by permission of Elsevier Ref. [2]

behavior similar to the undoped sample. The doped material reported by Chung et al. showed a lower capacity fading with the increase of the current. The better result exhibited by the doped $LiFePO_4$ could be partially ascribed to the low weight of the cathode, thereby requiring lower working currents (the 20C discharge rate corresponds to a current of 8.5 mA for the doped cathode and 30 mA for the cathode described in Fig. 8.8).

The reversibility of the material as a function of the cycle number is illustrated in Fig. 8.9. The cell was charged at a 1C rate, with a top-off at 4.0 V. This voltage was applied to the cells until the current was decreased to 1/10th of its initial value. The discharge step was driven at a 7C rate. The delivered capacity was about 50% of the theoretical capacity. Every 20 cycles, a cycle at reduced current density (C/3) was performed for testing the capacity under less stressed conditions. The specific capacity delivered after 200 discharge cycles at 50% depth of discharge (DOD) was as high as 156 Ah kg^{-1}. The capacity fade, evaluated by considering the capacity exhibited in the test cycles, was about 0.012% per cycle.

8.3 Conclusions

The electrochemical performance of undoped $LiFePO_4$ was found to be strictly dependent on both the amount of carbon in the composite cathodes and the procedure used to prepare the electrodes. The carbon mixing procedure influences both the electrode rate capability and the capacity fading upon cycling. For discharge currents lower than 3C rate, the electrochemical performance of well mixed, high carbon content, undoped $LiFePO_4$ is comparable with those of carbon coated and doped materials. At larger discharge rates, the high value the electrolyte, charge and mass transfer resistances drastically reduced the delivered energy. When discharged at 50% DOD, the electrode showed very interesting capacity retention with a capacity fade of only about 0.012% per cycle (fading evaluated by considering the capacity exhibited in low discharge current cycles). From these results it is possible to conclude that the electrochemical performance of undoped $LiFePO_4$ strongly depends upon the method of contacting the active material particles with the electronic conductor. For carbon coated $LiFePO_4$, the electronic contact between the grains is formed during the synthesis and the electric contact with the additional carbon added during the electrode preparation is easily established, enhancing the overall kinetics of the redox reaction. The electrochemical performance of doped $LiFePO_4$ with high electronic conductivity is almost unaffected by the discharge current, confirming that the electronic contact between the active material and the conductive filler is very good. In the case of undoped $LiFePO_4$, however, an extensive mixing procedure and high carbon content are required to achieve full active material utilization and low cycle fading. This aspect should be carefully evaluated when considering the material for practical applications. The optimization of powder mixing, i.e., by using a liquid-based method to mix carbon and active materials, may further decrease the amount of carbon needed to achieve the best electrochemical performance, increasing the overall cathode energy.

References

1. P.P. Prosini, M. Carewska, S. Scaccia et al., A new synthetic route for preparing $LiFePO_4$ with enhanced electrochemical performance. J. Electrochem. Soc. **149**, A886–A890 (2002)
2. D. Zane, M. Carewska, S. Scaccia et al., Factor affecting rate performance of undoped $LiFePO_4$. Electrochim. Acta **49**, 4259–4271 (2004)
3. M. Herstedt, M. Stjerndahl, A. Nytén et al., Surface chemistry of carbon-treated $LiFePO_4$ particles for Li-ion battery cathodes studied by PES. Electrochem. Solid St. **6**, A202–A206 (2003)
4. G.B. Appetecchi, M. Carewska, F. Alessandrini et al., Characterization of PEO-based composite cathodes—I Morphological, thermal, mechanical, and electrical properties. J. Electrochem. Soc. **147**, 451–459 (2000)
5. P.P. Prosini, S. Passerini, A lithium battery electrolyte based on gelled polyethylene oxide. Solid State Ionics **146**, 65–72 (2002)
6. S.-Y. Chung, J.T. Bloking, Y.-M. Chiang, Electronically conductive phospho-olivines as lithium storage electrodes. Nat. Mater. **1**, 123–128 (2002)
7. Z. Chen, J.R. Dahn, Reducing carbon in $LiFePO_4$/C composite electrodes to maximize specific energy, volumetric energy, and tap density. J. Electrochem. Soc. **149**, A1184–A1189 (2002)
8. H. Huang, S.-C. Yin, L.F. Nazar, Approaching theoretical capacity of $LiFePO_4$ at room temperature at high rates. Electrochem. Solid St. **4**, A170–A172 (2001)
9. E.M. Bauer, C. Bellitto, M. Pasquali et al., Versatile synthesis of carbon-rich $LiFePO_4$ enhancing its electrochemical properties. Electrochem. Solid St. **7**, A85–A87 (2004)

Chapter 9
Versatile Synthesis of Carbon-Rich $LiFePO_4$

9.1 Preparation of Nano-Particle $LiFePO_4/C$ Composites

Many authors have attempted to improve the electrochemical performance of $LiFePO_4$ by coating the surface with conducting particles [1], or cosynthesizing the compound with conductive additives [2–3]. In both cases the conductive particles interfered with the grain coalescence determining the reduction of the grain size. In this chapter we stress this concept reporting on a new reproducible synthetic route to prepare nano-particle $LiFePO_4/C$ composites, in which the phosphorus, iron and carbon atoms all originate from the same precursor. $LiFePO_4/C$ composites were prepared from thermal decomposition of Fe(II)organo-phosphonates $Fe[(RPO_3)(H_2O)]$ (R = methyl or phenyl group) in presence of Li_2CO_3 at high temperature and under inert atmosphere [4]. $Fe[CH_3PO_3]\cdot H_2O$ and $Fe[C_6H_5PO_3]\cdot H_2O$ have been synthesized as previously reported [5, 6]. The compounds are stable to the air and moisture. Li_2CO_3 and $Fe[(RPO_3)(H_2O)]$, (R = CH_3-, C_6H_5-) were mixed together in the ratio 1:2 in air to get an intimate mixture without any oxidation of Fe(II). Crystalline $LiFePO_4$ samples were obtained by heating the precursors in a tubular furnace under inert atmosphere of N_2 gas at temperatures >600°C for at least 16 h. This method provides $LiFePO_4$ samples of nano-particle size. X-ray diffraction (XRD) and Scanning Electron Microscope (SEM) were used for the characterization of the powders.

Composite cathode tapes were made by roll milling a mixture of 75 wt% active material and 10 wt% binder (Teflon, DuPont). Carbon (KJB Carbon) was added to have a 15 wt% final carbon content. Electrodes were punched in form of discs typically with a diameter of 10 mm. The electrode weight ranged from 7.4 to 10.7 mg. Electrochemical characterization of $LiFePO_4$ was performed in T-shaped battery cells with lithium metal as a counter and a reference electrode. The cells were filled with a 1M solution of $LiPF_6$ in ethylene carbonate/diethyl carbonate (1:1). The cycling tests were carried out automatically by means of a battery cycler (Maccor 4000). Composite cathode preparation, cell assembly, tests, and storage were performed in the dry room (R.H. < 0.1% at 20°C).

P. P. Prosini, *Iron Phosphate Materials as Cathodes for Lithium Batteries*,
DOI: 10.1007/978-0-85729-745-7_9, © Springer-Verlag London Limited 2011

Fig. 9.1 SEM micrograph of the $LiFePO_4$ sample, as obtained from the decomposition of $Fe[(C_6H_5PO_3)(H_2O)]$. Reproduced by permission of The Electrochemical Society Ref. [4]

9.2 Physical and Electrochemical Characterization

The TG curves of mixtures of lithium carbonate with Fe(II) methyl- and phenylphosphonate showed a weight loss over the temperature range from ambient to 200°C corresponding to the elimination of water molecules coordinated to the Fe(II) ions. Further weight loss of $\approx$30 and 45%, respectively was observed in the temperature region 200–800°C. The DTA curves showed a exothermic effect at 400°C for the $Fe[(CH_3PO_3)(H_2O)]$ and at 450°C for the $Fe[(C_6H_5PO_3)(H_2O)]$. These effects are probably related to the decomposition of carbonate to Li_2O followed by the formation of $LiFePO_4$. The exact mechanism of decomposition is not yet clear [7]. Crystalline $LiFePO_4$ samples were obtained by heating the precursors in a tubular furnace under inert atmosphere of N_2 gas at temperatures $>$600°C for at least 16 h. Elemental carbon is formed during the decomposition of the Fe(II) methyl- and phenyl-phosphonates, and it was found in the final samples in an amount of 2.5 and 12 wt%, respectively. Considering that the theoretical carbon content in the starting phosphonate materials corresponds to 8.8 and 29.6 wt% it follows that part of carbon is lost during the firing process. Figure 9.1 shows a SEM micrograph of the $LiFePO_4$ sample as obtained from the decomposition of $Fe[C_6H_5PO_3)(H_2O)]$. The $LiFePO_4$ phase consists of spherical aggregates of about 0.2 μm diameter. The SEM micrograph of the sample obtained using the $Fe[(CH_3PO_3)(H_2O)]$ shows a similar grain structure. In both cases the grain size has an average diameter $<$1 μm. The X-ray powder diffraction patterns of the compound were indexed in the orthorhombic space group *Pnma* and the unit-cell parameters are reported in Table 9.1. Rietveld refinement was performed on a model based on the single-crystal structure of $LiFePO_4$ (olivine structure). The similarity between the unit-cell parameters found in the material and those reported in the literature indicates the presence of olivine $LiFePO_4$.

Table 9.1 Rietveld refinement for $LiFePO_4$ in Pnma space group

a (Å)	b (Å)	c (Å)	X^2
10.325	6.006	4.691	2.1

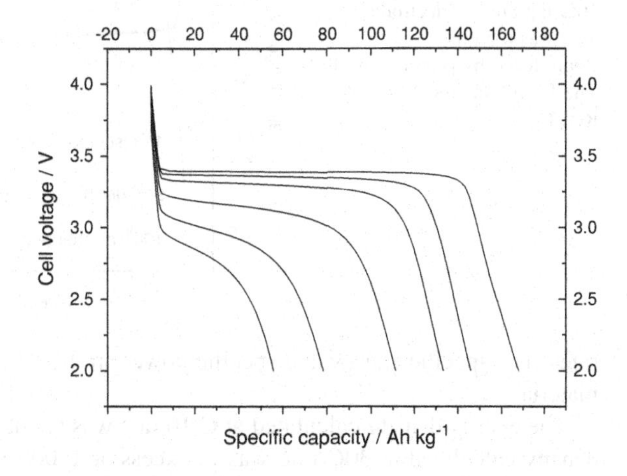

Fig. 9.2 Voltage profiles for the cell discharged at different rates. $LiFePO_4$ was prepared starting from the Fe(II) phenylphosphonate. The cathode loading of $LiFePO_4$ was 7.1 mg cm^{-2}. Electrode area was 0.16 cm^2. The cell was discharged at C/10, C, 3C, 10C, 20C, and 30C

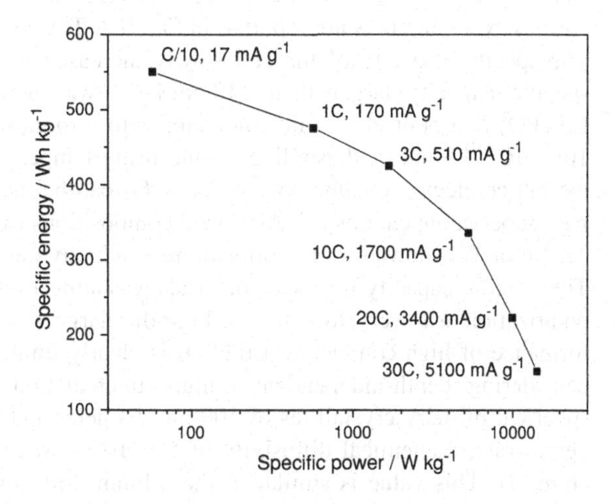

Fig. 9.3 Ragone plot for the cell discharged at different rates. The experimental conditions are the same as Fig. 9.2. Reproduced by permission of The Electrochemical Society Ref. [4]

The material was used as cathode in non-aqueous lithium cells. The cells were subjected to various discharge rates, i.e. C/10, 1C, 3C, 10C, 20C, and 30C. The cells were always charged using the same procedure to ensure identical initial conditions: a constant current step at 1C rate until the voltage reached 4.0 V, followed by a constant voltage step until the current fell below C/10 rate. Figure 9.2 shows the discharge voltage profiles. At C/10 discharge rate the full capacity is obtained (about 170 Ah kg^{-1}). By increasing the discharge current the capacity decreased progressively and at 30C rate the capacity was about 58 Ah kg^{-1}. Figure 9.3 shows the Ragone plot for the cell discharged at different

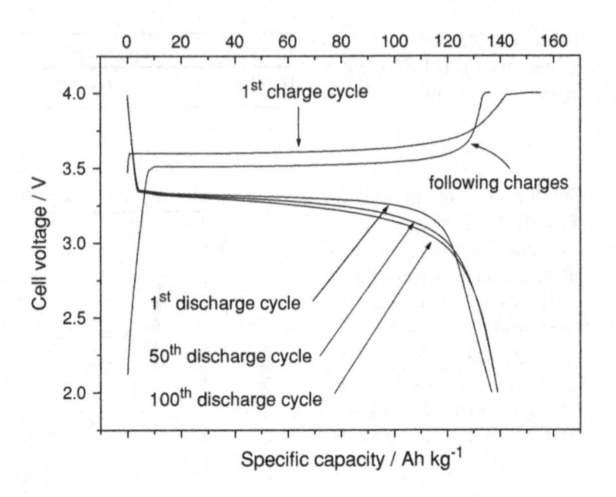

Fig. 9.4 Voltage profiles recorded during the first and following charge–discharge cycles. The cathode loading of $LiFePO_4$ was 10.2 mg cm^{-2}. Electrode area was 0.5 cm^2. Reproduced by permission of The Electrochemical Society Ref. [4]

rates. The specific energy and specific power are based on the weight of the active material.

The energy density calculated at C/10 rate was about 550 Wh kg^{-1}. The power density calculated at 30C rate was in excess at 14000 W kg^{-1} while the specific energy was about 28% of the energy delivered at C/10. The voltage profiles at various cycle numbers are reported in Fig. 9.4. The cell was discharged at 1C rate. The specific capacity of the cell slowly increased with the cycle number and a specific capacity larger than 140 Ah kg^{-1} was delivered during 100 cycles. $LiFePO_4$ is a poor electronic conductor with a room-temperature conductivity of 10^{-9}–10^{-10} S cm^{-1}, depending on the firing temperature [8]. It was shown that the lattice electric conductivity of $LiFePO_4$ can be increased by selective doping with supervalent cations [7]. All doped compositions $Li_{1-x}M_xFePO_4$ (M = Mg, Al, Ti, Nb or W) showed room-temperature conductivities in excess of 10^{-3} S cm^{-1}. The specific capacity for these materials was almost rate independent and the cell polarization was very low also at high-discharge rate. The electrochemical performance of high-conductive $LiFePO_4$ is clearly limited by lithium transport. By considering depth-independent lithium intercalation through the electrode of spherical primary crystallites of 100 nm diameter and using the data for capacity versus rate, a chemical diffusivity of 5×10^{-15} was calculated at room temperature [7]. This value is similar to the lithium diffusion coefficient calculated for un-doped materials [9, 10].

From this result it is possible to say that the lithium ions diffusion in $LiFePO_4$ is intrinsically low and that it does not depend on the component diffusion coefficient of electrons. This affirmation is also supported by the results described in this work. The electrochemical performance of the un-doped $LiFePO_4$ was found similar, if not better, than that of the corresponding highly-conductive material.

The outstanding performance of the $LiFePO_4$/C composites can be ascribed to the tailored synthesis addressed to enhance the electrochemical properties of the material. During the synthesis of $LiFePO_4$ part of the organic constituent of the precursor is oxidized to form elemental carbon. The carbon particles interact with the $LiFePO_4$ grains just during their formation interfering with the grains coalescence and addressing the grain size to nanometric dimension. Besides, the carbon particles present on the $LiFePO_4$ grain surface provides a good electronic contact between the grains, and the carbon added for the composite electrode fabrication decreases the charge transfer resistance. The low-particle size and the enhanced surface conductivity both increase the electrochemical performance of the material when used as a cathode in lithium batteries.

9.3 Conclusions

A new synthetic route leading to a $LiFePO_4$/C composite with very attractive electrochemical properties was proposed. The synthetic way suggested is simple and the organo-phosphonate used as precursors are very easy to prepare and stable to the air. Electrodes prepared with the $LiFePO_4$/C synthesized from Fe(II) phenyl phosphonate showed very impressive specific energy, specific power, and capacity retention upon cycling. The excellent electrochemical performance of the material makes the synthetic route a promising way to prepare a cathode material for fabrication of high power high energy lithium-ion batteries.

References

1. N. Ravet, J.B. Goodenough, S. Besner et al., Improved iron based cathode material. In *Proceeding of 196th ECS Meeting*, Hawaii, 17–22 Oct 1999
2. H. Huang, S.-C. Yin, L.F. Nazar, Approaching theoretical capacity of $LiFePO_4$ at room temperature at high rates. Electrochem. Solid St. **4**, A170–A172 (2001)
3. P.P. Prosini, D. Zane, M. Pasquali, Improved electrochemical performance of a $LiFePO_4$-based composite cathode. Electrochim. Acta **46**, 3517–3523 (2001)
4. E.M. Bauer, C. Bellitto, M. Pasquali et al., Versatile synthesis of carbon-rich $LiFePO_4$ enhancing its electrochemical properties. Electrochem. Solid St. **7**, A85–A87 (2004)
5. C. Bellitto, F. Federici, M. Colapietro et al., X-ray single-crystal structure and magnetic properties of $Fe[CH_3PO_3)]\bullet H_2O$: A layered weak ferromagnet. Inorg. Chem. **41**, 709–714 (2002)
6. A. Altomare, C. Bellitto, S.A. Ibrahim et al., Synthesis, X-ray powder structure, and magnetic properties of the new, weak ferromagnet iron(II) phenylphosphonate. Inorg. Chem. **39**, 1803–1808 (2000)
7. E.M. Bauer, C. Bellitto, G. Righini et al., A versatile method of preparation of carbon-rich $LiFePO_4$: A promising cathode material for Li-ion batteries. J. Power Sources **146**, 544–549 (2005)
8. S.-Y. Chung, J.T. Bloking, Y.-M. Chiang, Electronically conductive phospho-olivines as lithium storage electrodes. Nat. Mater. **1**, 123–128 (2002)
9. P.P. Prosini, M. Lisi, D. Zane et al., Determination of the chemical diffusion coefficient of lithium in $LiFePO_4$. Solid State Ionics **148**, 45–51 (2002)
10. S. Franger, F. Le Cras, C. Bourbon et al., $LiFePO_4$ synthesis routes for enhanced electrochemical performance. Electrochem. Solid St. **5**, A231–A233 (2002)

Chapter 10
Modeling the Voltage Profile for $LiFePO_4$

10.1 Experimental

Crystalline $LiFePO_4$ was obtained by heating amorphous $LiFePO_4$ in a tubular furnace at 550°C under a reducing atmosphere ($Ar/H_2 = 95/5$) for 1 h. The active material and carbon (KJB, Akzo Nobel) were mixed in a mortar for 15 min. The binder (Teflon, DuPont) was added and the blend was mixed to obtain a plastic-like material. The quantity of binder and carbon were kept constant at 10 and 20 wt%, respectively. Composite cathode tapes were made by roll milling the mixture. Electrodes were punched in the form of discs, typically with a diameter of 8 mm. Polypropylene *T*-type pipe connectors with three cylindrical stainless steel (SS316) current collectors were used as cells. Lithium was used both as an anode and a reference electrode and a glass fibre was used as a separator. The cells were filled with a 1M solution of $LiPF_6$ in ethylene carbonate/diethyl carbonate (1/1). The cycling tests were automated using a battery cycler (Maccor 4000). Composite cathode preparation, cell assembly, tests, and storage were performed in a dry room (R.H. < 0.1% at 20°C).

10.2 Modeling the Voltage Profile

To explain the lithium insertion/deinsertion in $LiFePO_4$, Padhi et al. [1] proposed that the lithium motion proceeds from the surface of the particle moving inwards through a two-phase interface (shrinking core model). Andersson et al. in addition to the "radial model" [2] proposed a "mosaic model" [3] that invokes a mosaic character within each particle. More recently Newman studied the lithium insertion in $LiFePO_4$ using the shrinking core model [4]. Upon charge a Li-rich core is covered by a Li-deficient shell while the Li-rich shell is formed on the Li-deficient core upon discharge. Delmas et al. [5] proposed a "domino-cascade model" in which the existence of structural constraints, occurring just at the reaction

P. P. Prosini, *Iron Phosphate Materials as Cathodes for Lithium Batteries*,
DOI: 10.1007/978-0-85729-745-7_10,

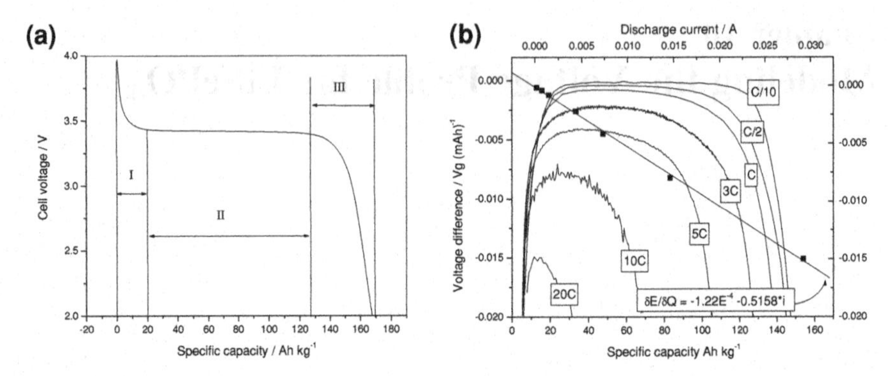

Fig. 10.1 **a** Discharge voltage profile for a cell discharged at a C/10 rate. Specific charge current: 17 A kg^{-1}. Cathode loading of $LiFePO_4$: 10.9 mg cm^{-2}. **b** Differential voltage as a function of the intercalation degree at various discharge currents (*bottom*). The voltage difference is also plotted as a function of the discharge current (*top*). The equation is the linear fit of this latter curve from which the R_{mt} was evaluated ($R_{mt} = 0.5158\ \Omega$ kg $(Ah)^{-1}$). Reproduced by permission of The Electrochemical Society Ref. [5]

interface, lead to the minimization of the elastic energy thus enhancing the deintercalation (intercalation) process. In this chapter, starting from the claim that by growing from the centre the delithiated phase can reduce the stress originating from volume contraction, a general equation describing the voltage profile as a function of the intercalation degree will be developed as a function of the discharge rate [6]. It is well known from the Gibbs rule that phase segregation can result as a consequence of energy minimization. The newly formed phase can segregate on the grain surface. In this case the new phase starts to grow from the edge and moves toward the center. Otherwise, if the parent phase segregates on the grain boundary, the new phase starts to grow from the center and moves toward the edges. In the case of $LiFePO_4$, the strong interactions between the lithiated and delithiated phases tend to push the latter inside the grains, with the parent phase segregated to the grain boundary. The force driving the accommodation of the delithiated phase inside the grains can be related to volume variations. By growing from the center, the delithiated phase can reduce the stress originating from volume contraction. If the new phase starts to grow from the grain edge it should occupy a volume smaller than the parent phase, increasing the pressure between the phase boundary. If the volume variation is negative (the volume of the new phase is smaller than the volume of the parent one, as in the case of $FePO_4$ growing in $LiFePO_4$) the new phase starts to grow from the center. If the volume variation is positive, the new phase starts to grow from the edge. In this work it was assumed that, upon discharge, $LiFePO_4$ starts to grow from the edge and, moving inside the grain, tends to reach the center.

Figure 10.1a shows the cell voltage profile upon discharge for a cell discharged at C/10 rate. The voltage profile of $LiFePO_4$ rapidly declines from the end charge voltage to about 3.42 V versus Li(I). This voltage remains practically unchanged upon further lithium intercalation (II). Near the end of the discharge, the voltage profile starts to decrease and rapidly reaches the end discharge voltage (III). Upon lithium insertion, grain boundary segregation causes a rapid variation of the lithiated phase on the grain surface resulting in a sharp change in the cell voltage. The cell voltage stops changing after the insertion of about 20 Ah kg^{-1} (corresponding to $x = 0.1$ in $Li_{1-x}FePO_4$). The invariance of the cell voltage upon further lithium insertion enables us to state that, at this stage, the lithiated phase completely covers the grain surface. Assuming that the material is formed of regular spheres of about 100 nm in diameter, the composition $x = 0.1$ corresponds to the filling of few atomic layers on the grain surface. The penetration depth (*d*) of the lithiated phase as a function of the intercalation degree (*x*) and the grain radius (*r*) is $d = r[1-(1-x)^{1/3}]$. For $r = 50$ nm and $x = 0.1$ the penetration of the lithiated phase is about 17 Å which roughly corresponds to the *c* parameter or to one half of the parameters *a* and *b* of the unitary cell.

The logistic Eq. 10.1 was used to describe the voltage behavior:

$$E = \left(E_{ec} - E_1^\circ\right)/\left(1 + x/x_1^\circ\right) + E_1^\circ \tag{10.1}$$

where E_{ec} is the end charge voltage, E°_1 is the plateau voltage, x is the intercalation degree and x°_1 is the composition for which the voltage is one-half of the voltage difference ($E_{ec} - E^\circ$). By fitting the voltage profile with Eq. 10.1, x° and E°_1 are found to be 2.0 Ah kg^{-1} and 3.40 V, respectively.

Upon further lithium insertion, the cell voltage remains almost independent from the intercalation degree due to the fact that the new phase grows inside the grain leaving the surface composition unchanged. By observing the voltage profile in the flat region, especially for higher discharge currents (see for example Fig. 10.7), it is possible to note a deviation of the voltage profile from the expected behavior, namely a decrease of the average discharge voltage and the presence of a voltage slope.

The decrease of the average discharge voltage can be related to the ohmic drop by means of the electrolyte resistance (R_e) and charge transfer resistance (R_{ct}), while the slope in the voltage profiles can be related to a progressive resistance to incorporate lithium which is dependent upon the intercalation degree (mass transfer resistance R_{mt}). From Ohm's law, the voltage drop is:

$$\Delta E = i \cdot [(R_e + R_{ct}) + x \cdot R_{mt})] \tag{10.2}$$

where i is the discharge current and x is the intercalation degree. To evaluate the different contributes to the cell voltage, R_{mt} was first evaluated. Figure 10.1b shows the differential voltage as a function of the intercalation degree at various discharge currents.

The voltage variations due to the R_{mt} increase by increasing the discharge current. The voltage drop in the flat region was calculated taking into account most of the lithium insertion process. The voltage drop is reported in the same figure as a function of the discharge current. A linear behavior is observed with a curve slope from which R_{mt} was evaluated ($R_{mt} = 0.5158\ \Omega$ kg $(Ah)^{-1}$). The R_{mt} evaluated for the full charged material was 87.6 Ω. After removing the R_{mt}

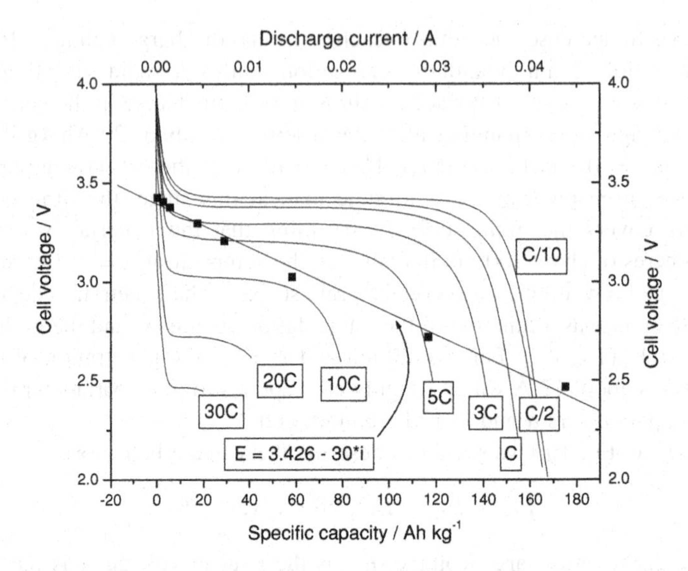

Fig. 10.2 Voltage profiles as obtained after removing the R_{mt} contribution as a function of the intercalation degree at various discharge currents (*bottom*). The voltage drop is also plotted as a function of the discharge current (*top*). The equation is the linear fit of this latter curve from which the sum of R_e and R_{ct} was evaluated to be 30 Ω. Reproduced by permission of The Electrochemical Society Ref. [5]

contribution to the voltage profiles, they appear as plotted in Fig. 10.2. The flat region is clearly more pronounced and it is now possible to easily evaluate the voltage drop (the values are plotted in the same figure as a function of the discharge current). For currents lower than 20C, a linear behavior was observed. From the curve slope, the sum of R_e and R_{ct} was evaluated to be 30 Ω.

Figure 10.3 is a pictorial description of the mechanism of lithium insertion/deinsertion in the system $LiFePO_4/FePO_4$. In (1) a grain of $LiFePO_4$ in which $FePO_4$ is growing is depicted. The increase of the voltage difference applied to the electrode forces an electron to be removed from the grain surface (2).

The carbon particle on the grain surface allows the electron to reach the current collector. On the grain surface a hole is formed as a consequence of the voltage increase. If the imposed current is very high lithium is extracted from the surface increasing the cell voltage. If the hole lifetime is larger enough to allow the hole to move inside the material, the hole can migrate from the grain surface to the $FePO_4/LiFePO_4$ interface (3). The lithium ion at the $FePO_4/LiFePO_4$ interface is surrounded by the $FePO_4$ phase and is in a high energy status. It can relax by pushing all of the lithium ions in the same column (5). At the end of the process, a lithium ion is deinserted from the edge and the new phase is formed at the $FePO_4/LiFePO_4$ interface (6). The same mechanism can be evoked for the lithium insertion process. Starting from (6), the decrease of the cell voltage forces a lithium ion to be inserted in the structure (5). The insertion of the lithium ion pushes all of the lithium ions in the same column and the hole is formed at the $FePO_4/LiFePO_4$

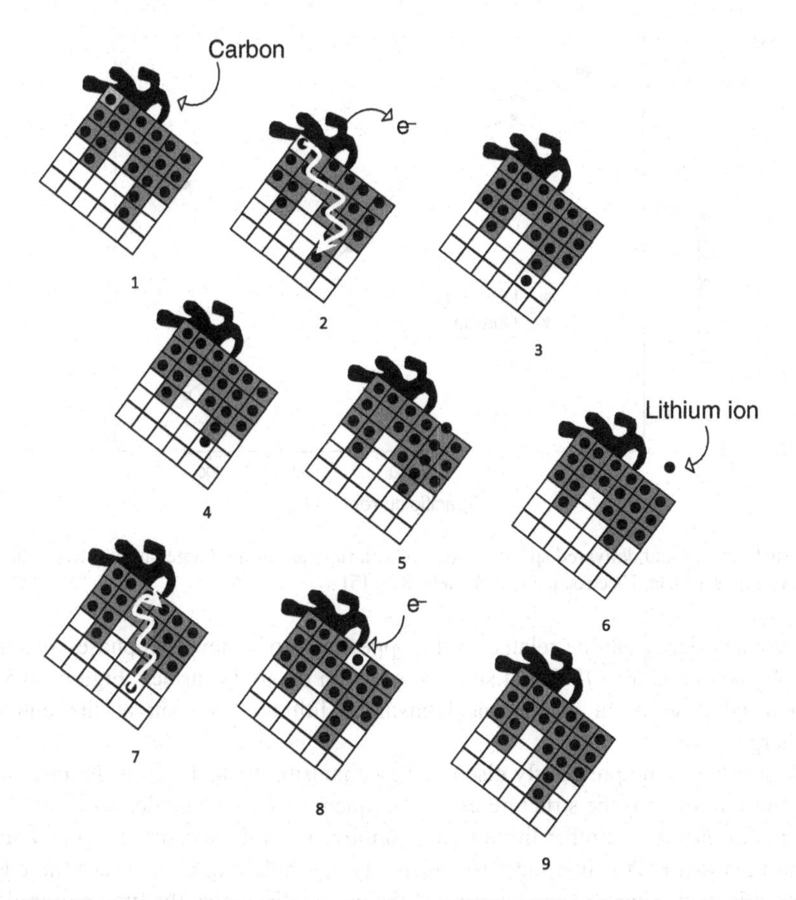

Fig. 10.3 Pictorial description of the mechanism of lithium (*black circle*) insertion/deinsertion in $LiFePO_4$ (*dark square*) and $FePO_4$ (*white square*). See the text for explanation. Reproduced by permission of The Electrochemical Society Ref. [5]

interface (4). The voltage difference applied to the electrode drives the hole to migrate from the center to the edge (7) and when it arrives near the grain surface, an electron can be accepted (8) restabilising the electroneutrality and stabilizing the structure (9). The R_{mt} increases linearly with the intercalation degree since the number of lithium ions that should be moved to accommodate the lithium insertion increases. Upon charge a similar limitation mechanism is not effective since the R_{mt} should decrease by decreasing the intercalation degree. Figure 10.4 shows the stored and delivered specific capacity as a function of discharge current.

Independently from the rate, the stored capacity is larger than the delivered one confirming that a different limitation mechanism works upon charge (i.e., the formation of delithiated phase directly on the grain surface). Referring to Fig. 10.1a, after insertion of 135 Ah kg^{-1} ($Li_{0.8}FePO_4$) and for further lithium insertion, the voltage profile rapidly decreased to reach the end charge voltage.

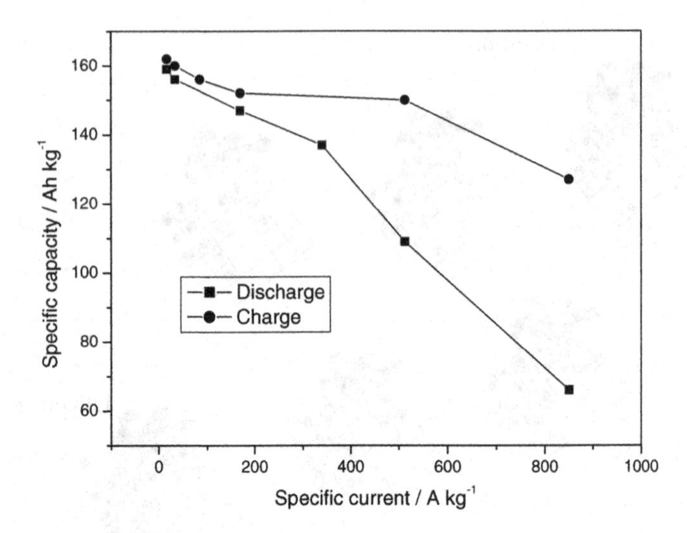

Fig. 10.4 Stored and delivered specific capacity as a function of discharge current. Reproduced by permission of The Electrochemical Society Ref. [5]

The voltage decay can be related to the appearance of a new two-phases system, namely the couple Fe^{2+}/Fe° that starts to grow on the grain surface. Figure 10.5 is a pictorial description of the mechanism of lithium insertion at the end of discharge.

According to the previously discussed mechanism, the first step is the insertion of a lithium ion into the structure as a consequence of a voltage decrease (a). The lithium ion pushes the other lithium ions to move toward the center leaving a hole at the $FePO_4/LiFePO_4$ interface (b). Normally, the hole migrates toward the edge where it is neutralized by an electron. At the end of discharge, the time required to reach the edge increases due to the increase of the lithiated phase and to the interface shrinkage (which increases the current density). A critical value is reached when the number of lithium ions inserted becomes larger than the number of holes that can be formed and an electron must be injected on the grain surface before the hole arrival (b). Since $LiFePO_4$ is a very poor electron conductor, the electron remains localized on the grain surface rather than moving to neutralize the incoming hole. The iron atom on the surface is in a hypothetical Fe^{+1} oxidation state. At this stage a second lithium ion is inserted on the grain surface neutralizing the negative charge (c). Now, a second electron can be injected at the grain surface (d), forming iron metal (e), and a third lithium ion inserted restabilising the electro-neutrality and forming lithium phosphate on the grain surface (f). Finally, the hole reaches the grain surface and a third electron is injected to neutralize the hole. At the end of the process on the grain surface will appear iron metal and lithium phosphate (g).

To simulate the cell voltage during stage (III), the sigmoid function was used:

$$E = (E_1^{\circ} - E_2^{\circ})/(1 + \exp((x - x_2^{\circ})/\mathrm{d}x) + E_2^{\circ} \tag{10.3}$$

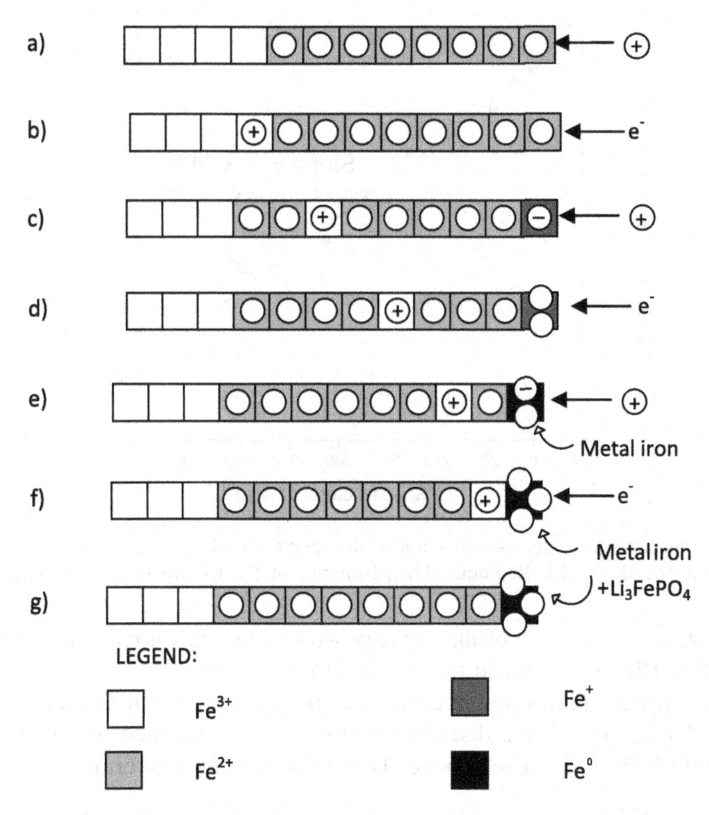

Fig. 10.5 Pictorial description of the mechanism of lithium insertion at the end of discharge. Lithium (*circle*), $LiFePO_4$ (*dark square*), $FePO_4$ (*white square*). See the text and the legend for explanation. Reproduced by permission of The Electrochemical Society Ref. [5]

where $E°_1$ and $E°_2$ are the equilibrium voltage for the couples Fe^{3+}/Fe^{2+} and $Fe^{2+}/Fe°$, respectively, x is the intercalation degree, $x°_2$ is the composition for which the voltage is one half of the ($E°_1 - E°_2$) voltage difference and dx takes into account the rapidity of the voltage change in proximity of $x = x°$. To a first approximation, the end charge voltage can be considered as one half of the ($E_{ec} - E°$) voltage difference. In such a case, $x°_2$ corresponds to the discharged capacity.

The discharged capacity was a function of the discharged current.

The capacity decline with increasing currents can be interpreted taking into account the same mechanism responsible for the sharp voltage decay at the end of the discharge of the low current cycle, namely the appearance of iron metal on the grain surface. The intercalation degree for which iron metal starts to appear on the grain surface depends upon the applied current. To evaluate this dependence the discharge capacity is plotted in Fig. 10.6 as a function of the specific discharge current. A linear behavior was observed. For discharge currents lower than 20C, the discharged capacity ($x°$) is: $x° = Q° + ki$ where $Q°$ is the theoretical specific capacity for $LiFePO_4$ (170 Ah kg^{-1}) and k is a constant (in hours) that takes into

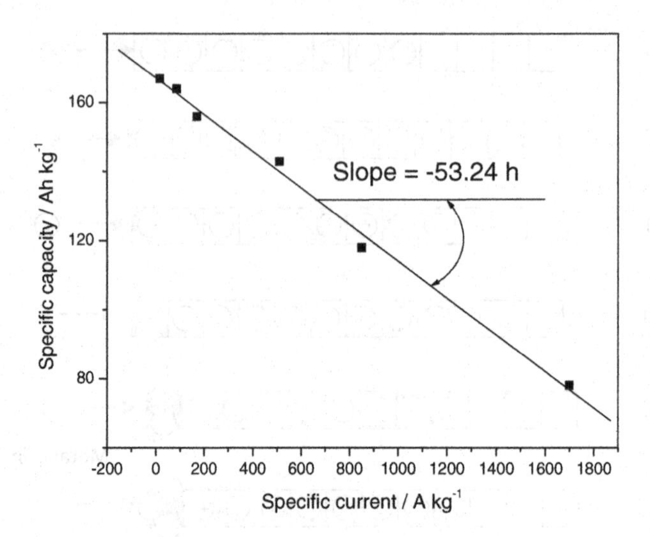

Fig. 10.6 Discharge capacity as a function of the specific discharge current. The slope of the linear fit was found −53.2 h. Reproduced by permission of The Electrochemical Society Ref. [5]

consideration the decrease of the capacity with increasing currents. By fitting the data in Fig. 10.6, k was found to be −53.24 h.

A general equation to describe the voltage profile as a function of the intercalation degree (x) and the discharge current (i) was obtained by summing the single contributions discussed above. The equation takes the form:

$$E = \left(E_{ec} - E_1^\circ\right)/(1 + x/x^\circ) + \left(E_1^\circ - E_2^\circ\right)/ \\ \left(1 + \exp((x - (Q^\circ - \mathrm{ki/kg}))/\mathrm{d}x\right) + E_2^\circ - i \cdot \left[(R_e + R_{ct}) + x \cdot R_{mt})\right] \tag{10.4}$$

Table 10.1 reports the values of the parameters obtained fitting the C/10 discharge curve by using Eq. 10.4.

Figure 10.7 shows the results obtained by using Eq. 10.4 to fit the voltage profiles at various discharge currents. The proposed equation could be used to evaluate the state of charge of the material by fitting the discharge curve with an appropriate algorithm. Fongy et al. [7] used the proposed model to extract from the $LiFePO_4$ discharge curves two parameters. The parameters were employed to determine the optimal electrode engineering and to interpret the origins of the electrode performance limitations.

10.3 Conclusions

A general equation to describe the discharge voltage profiles for $LiFePO_4$ was developed. The equation was based on a model requiring phase segregation, with

Table 10.1 Values of the parameters obtained fitting the C/10 discharge curve and using Eq. 10.4. Reproduced by permission of The Electrochemical Society Ref. [5]

Description	Symbol	Value
End charge voltage	E_{ec}	4.00 V
First voltage plateau	E°_1	3.40 V
Second voltage plateau	E°_2	1.06 V
Composition for which the voltage is ½($E_{ec}-E^{\circ}_1$)	x°_1	2.0 Ah kg^{-1}
Composition for which the voltage is ½($E^{\circ}_1-E^{\circ}_2$)	x°_2	164.0 Ah kg^{-1}
Composition for voltage decay at the end of discharge	dx	7.0 Ah kg^{-1}
Electrolyte and charge transfer resistance	$(R_e + R_{ct})$	30 Ω
Mass transfer resistance	R_{mt}	0.5158 Ω kg $(Ah)^{-1}$
Theoretical specific capacity	$Q°$	170 Ah kg^{-1}
Slope of capacity decrease	k	−53.2 h
Active material mass	kg	8.5×10^{-6} kg

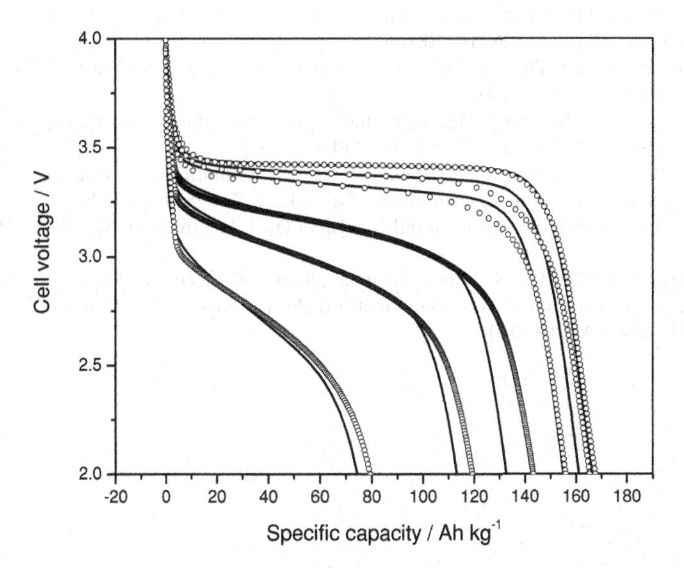

Fig. 10.7 Discharge voltage profiles for a cell prepared with the cathode described in Fig. 10.1. The cell was discharged at various rates ranging from C/10 up to 10C (*solid line*). The simulated voltage profiles as obtained from Eq. 10.4 are also reported (*circle*). Reproduced by permission of The Electrochemical Society Ref. [5]

the delithiated phase inside the grain, and with electronic diffusion lower than ionic diffusion. The slow electronic diffusion was responsible for the decrease of the electrochemical performance of the material with increasing currents. The fall in the voltage profile near the end of the discharge was related to the formation of lower valence iron on the grain surface. The simulation fitted very well the real curve at the lowest discharge rate where the parameters were calculated, but it was also valid for discharge currents as high as 10C. The proposed equation could be used to evaluate the state of charge of the material by fitting the discharge curve with an appropriate algorithm. Since the adjustable parameters reflect macroscopic and microscopic variables such as grain size and shape (the parameter $x°1$, dx and R_{mt}), electrode porosity (the parameter k), electrolyte solvent and salt (the parameter $R_e + R_{ct}$), the model can be applied to simulate the voltage profiles for materials with different grain size and morphology or with different electrode porosity or for cell prepared with different electrolyte. The adjustable parameters have to be evaluated following the above described procedure and the equation used to fit the voltage profiles as a function of the discharge current.

References

1. A.K. Padhi, K.S. Nanjundaswamy, J.B. Goodenough, Phospho-olivines as positive-electrode materials for rechargeable lithium batteries. J. Electrochem. Soc. **144**, 1188–1194 (1997)
2. A.S. Andersson, J.O. Thomas, B. Kalska et al., Thermal stability of $LiFePO_4$-based cathodes. Electrochem. Solid St. **3**, 66–68 (2000)
3. A.S. Andersson, J.O. Thomas, The source of first-cycle capacity loss in $LiFePO_4$. J. Power Sources **97–98**, 498–502 (2001)
4. V. Srinivasan, J. Newman, Discharge model for the lithium iron-phosphate electrode. J. Electrochem. Soc. **151**, A1517–A1529 (2004)
5. C. Delmas, M. Maccario, L. Crogunnec et al., Lithium deintercalation in $LiFePO_4$ nanoparticles via a domino-cascade model. Nat. Mater. **7**, 665–671 (2008)
6. P.P. Prosini, Modeling the voltage profile for $LiFePO_4$. J. Electrochem. Soc. **152**, A1925–1929 (2005)
7. C. Fongy, A.-C. Gaillot, S. Jouanneau et al., Ionic vs Electronic Power Limitations and Analysis of the Fraction of Wired Grains in $LiFePO_4$ Composite Electrodes. J. Electrochem. Soc. **157**, A885–A891 (2010)

Index

P. P. Prosini, *Iron Phosphate Materials as Cathodes for Lithium Batteries*,
DOI: 10.1007/978-0-85729-745-7,